The Philosophical Foundations of Modern Medicine

The Philosophical Foundations of Modern Medicine

Keekok Lee
University of Manchester, UK

palgrave
macmillan

First published 2012 by
PALGRAVE MACMILLAN

Palgrave Macmillan in the UK is an imprint of Macmillan Publishers Limited, registered in England, company number 785998, of Houndmills, Basingstoke, Hampshire RG21 6XS.

Palgrave Macmillan in the US is a division of St Martin's Press LLC, 175 Fifth Avenue, New York, NY 10010.

Palgrave Macmillan is the global academic imprint of the above companies and has companies and representatives throughout the world.

Palgrave® and Macmillan® are registered trademarks in the United States, the United Kingdom, Europe and other countries.

ISBN: 978–0–230–34829–5

This book is printed on paper suitable for recycling and made from fully managed and sustained forest sources. Logging, pulping and manufacturing processes are expected to conform to the environmental regulations of the country of origin.

A catalogue record for this book is available from the British Library.

Library of Congress Cataloging-in-Publication Data

Lee, Keekok, 1938–
    The philosophical foundations of modern medicine / Keekok Lee.
        p. cm.
    Includes bibliographical references and index.
    ISBN 978–0–230–34829–5 (alk. paper)
        I. Title. [DNLM: 1. Philosophy, Medical. 2. Biomedical Engineering – history.
    W 61] LC-classification not assigned

610.28—dc23   2011029576

10 9 8 7 6 5 4 3 2 1
21 20 19 18 17 16 15 14 13 12

Transferred to Digital Printing in 2012

Contents

v

Acknowledgements

Naturally, as a mere philosopher and lay person writing about medicine, I have no confidence that I would not have made gross errors in my understanding of such a vast and complicated subject. Fortunately, I managed to persuade two people who very kindly consented to go through the medical aspects of the book. I am deeply grateful to Dr Leela Joseph, Retired Consultant Microbiologist, UK, who read and commented on some of the relevant sections as well as to Dr Ian Balfour, Retired Consultant Haematologist, UK, who very gallantly read and commented on the entire text. Indeed, their comments have saved me from numerous mistakes, great and small. However, I alone bear responsibility for any which may still remain, as I may have misunderstood their suggestions for change and emendation. However, I must make clear that they are not in any way responsible for the philosophical interpretation of the medicine explored in this book.

Introduction

This book about the philosophy of modern medicine[1] is written within the broad parameters of the framework as set out below:

1. Such a medicine is scientific medicine; as such it can only be understood as part of modern science.
2. The beginnings of such a medicine[2] may be dated to at least the seventeenth century just as modern science itself may similarly be dated.
3. Modern science cannot be understood in a vacuum without tracing it back to modern philosophy in which it is embedded.

The book explores in detail the implications of the three theses outlined above, establishing that the major characteristics of such a medicine as well as such a science follow more or less directly from their philosophical foundation and source.[3] Thus it is not an accident that modern medicine is atomistic, reductionist, mechanistic as well as technology-oriented as the philosophical worldview from which it follows is bounded by the same parameters.

Part I contains five chapters. These together set out the philosophical foundations of modern science as well as, therefore, modern medicine, in order to display why the latter possesses the features it does exhibit:

- Chapter 1 shows the intimate link between science and philosophy in general, and between modern science and modern philosophy in particular.
- Chapter 2 argues that every philosophy in which its science is embedded entails a methodology which that science follows – modern philosophy entails methodologically that modern science is objective/

1

quantifiable and reductionist in character. The imprimatur "science" in this tradition of science/philosophy would only be bestowed on data obtained using such methods.

- Chapter 3 explores, in the view of this author, one of the most radical philosophical ideas behind the modern scientific revolution. It is not the Copernican Revolution (whether narrowly or more broadly understood) but an ontological revolution – that is to say, the abandonment of the naturally-occurring mode of being for the artefactual mode of being.[4] This profound change was prepared by (amongst others) Descartes and his dualist thinking which opens a space for modern science, as its remit is confined only to matter (not soul/mind). Furthermore, organisms (whether non-human or human) are made of inert matter, subject to the laws of physics (and later other sciences such as chemistry) only. This is combined with the ontological *volte-face* that the organism is an artefact; more precisely, it is a particular type of artefact – a machine. Such a philosophical perspective is called mechanism, or the mechanistic world-view. Such a world-view entails reductionism: that the whole is no more than the sum of its parts, that once the parts have been explained, the whole has been explained without residue. At the same time, such a perspective has built into it the privileging of *homo faber* (over *homo cogitans*, Descartes notwithstanding) who manipulates, controls and transforms nature to suit their wishes and goals. Science is to provide the theoretical basis for generating technologies suitable for achieving this ideological goal, which prompts Heidegger to call science Theoretical Technology.
- Chapter 4 explores the notion of machines as Engineering, which will then show in greater detail why reductionism is entailed by the axiom that the body-is-machine.
- Chapter 5 examines in some detail the ontological *volte-face* of organism-is-machine by looking at the relationship between theoretical biology and philosophy as well as at the technologies engendered by the great discoveries of Mendelian genetics and molecular genetics/biology underpinning the two scientific agricultural/medical revolutions of the twentieth century.

Part II has three chapters which, in the light of Part I, explore in some detail the nature of modern medicine:

- Chapter 6 looks at, in general, the implications of the axiom that the body-is-machine for modern medicine. In particular, it looks at the implications of Engineering and engineering for Medicine and medicine.

- Chapter 7 shows that a temporal cleavage exists between basic medical sciences and therapies (mirroring a similar cleavage between theoretical science and technology in general); it argues why anatomy is the first medical science to be established, then followed by physiology. It also shows that medical technologies are necessarily and increasingly high tech in character, as they alone can give us more precise, more finely quantitative, more directly accessible data via machines.
- Chapter 8 demonstrates that medical technologies increasingly intervene at a deeper and deeper level of matter in tandem with the deeper and deeper levels of theoretical understanding of matter. It also looks at two specific forms of technological intervention, namely, surgery and pharmacology to show in particular that the former manifests in a more or less literal fashion that body-is-machine while showing that the latter displays the reductionist character in its various stages of development and design. Furthermore, it examines psychopharmacology to expose the precise philosophical framework within which it operates. It argues that the philosophical framework in question is epiphenomenalism (that matter can affect mind, but not mind matter). However, epiphenomenalism cannot make sense of the placebo effect; the most recent research shows that there is more to the placebo effect than meets the eye. This new understanding has resulted in the emergence of a new philosophical perspective which appears in turn to have the effect of challenging the philosophical as well as the methodological foundations of modern medicine itself.

Part III has four chapters:

- Chapter 9 looks at a sub-conception of the aetiological definition of disease, namely, the infectious-agent model of the monogenic approach. It looks at the reasons for its ascendancy since its emergence in the late nineteenth century, its continuing success as a progressive research programme (even today, a hundred years later); at the same time, however, it also looks at the anomalies which such a programme has to confront and the ways it has adopted to cope with them.
- Chapter 10 explores the causal model behind the infectious-agent monogenic conception to show that it is mono-factorial, linear and Humean in derivation. It sets out both its strengths and its

weaknesses. It distinguishes between three different contexts: (a) explanatory/scientific, (b) attributive, (c) clinical. From the first perspective, the chapter argues that no factor could be singled out as "the cause" as each of the relevant factors which may be identified, each on its own, is neither necessary nor sufficient – all the identified factors form a complex set of sufficient ("inus") conditions. On the other hand, from the second and third perspectives, it is legitimate for doctors to single out one of these "inus" conditions as "the cause". In the light of such a critical assessment, it is plausible to argue that the monogenic conception of disease be regarded as a methodological guideline in medical research about what factor(s) may count as cause(s) in diseases, rather than enunciating in a straight-forward fashion the empirical discovery of "the cause" of disease.

- Chapter 11 explores, in some detail, two attempts in the context of clinical medicine to articulate "the cause" of a disease, namely, the criterion of controllability/eliminability and the notion of the Random Controlled Trial (RCT). It argues that these two are closely related as the former's understanding of cause is implicated in the latter; that they both are involved in the notion of experiment; that Mill's methods, in the main, set out the logic of such experimentation; and that this sense of cause is what Collingwood calls Sense II.

- Chapter 12 examines a very different tradition, alongside the monogenic conception of disease, in the history of modern medicine which is embodied in the theory and practice of epidemiology. As its metaphysics is not atomistic but holist, its methodology is not reductionist; the notion of cause it deploys is multi-factorial and reciprocal or "ecosystemic". One could argue it is "revolutionary" science (whereas at the beginning of the twenty-first century, the infectious-agent model of disease may be said to be "normal" science). However, no Nobel award has been bestowed on the subject and its leading practitioner(s). This chapter attempts to make a case for saying that this may be a sadly-missed opportunity, as the "ecosystemic" kind of science shows signs of being the science of this new century.

Part I
Philosophy and Science

1
Philosophical Foundations

Science and philosophy

No intellectual activity, whether it pertains to politics, economics, law or science is innocent of philosophy, whatever its rhetorical proclamations may say. Hence to understand modern science as well as also modern medicine (of which it is a part), one must understand their philosophical foundations. However, before going any further, one must first address an important preliminary matter, namely, a terminological issue.

How is the term "science" understood? In its broadest sense, science is nothing but systematically organized knowledge which is how the German language uses the term *wissenshaft*. In English, ever since Francis Bacon (1561–1626) championed inductive reasoning as the basis of (modern) scientific method,[1] the term has been used characteristically in connection with the experimental physical or natural sciences, such as physics, chemistry, physiology, biochemistry, and so on. As a concession, the term has been extended to the social sciences, such as sociology and economics.[2] In centres of learning outside the English-speaking world, science as *wissenschaft* is unquestioned, such that historically, theology as well as Euclidean geometry (both being axiomatic systems[3]) were considered as paradigmatic sciences, while the physical/natural sciences are simply in the German tradition called *naturwissenschaft* and the social sciences *geisteswissenschaft*. This book proposes to follow the broader rather than the historically, narrower usage peculiar to the English-speaking world of science. It follows then, as we shall see, that there could be other systematic knowledges, all perfectly entitled to the label "science", such as ancient Greek science, (European) medieval science, just to mention a few examples. It also follows that each of these sciences is a consequence of their respective philosophies.[4]

Let us next consider how philosophy itself is to be understood. This is not a book about philosophy; as such, only the briefest account can be given here so that the link between on the one hand, philosophy and on the other, science can be understood. A crude, simplistic way of understanding the subject (though perhaps not the most satisfactory) is to look at the four main branches of philosophy, namely: metaphysics, epistemology, logic and ethics (values).

Take the last first, as it is the most familiar to lay people. Everyone is aware that one of the most fundamental philosophical preoccupations over the millennia is the attempt to answer the following questions: What is the good life? What is the good society? As individuals, how should we conduct ourselves towards others – to kith and kin as well as to neighbours and strangers? What criteria ought we to use to judge ourselves and others? Is a character which acts from good motives of greater moral significance or is it the good consequences of one's action which count for more than the motives/character of the agent? (A person who acts with the most laudable of motives may, nevertheless, end up doing something with really bad consequences for many, while someone who acts from a morally ignoble motive may end up doing something with truly beneficial consequences for a lot of other people.) Investigation into values covers not merely ethical values but also political, economic, aesthetic as well as environmental ones. Furthermore, as we shall see, it is also concerned with an important, though less familiar matter, to lay people which professional philosophers call the logic of ethical discourse or meta-ethics – we shall be referring to this briefly a little later. But in recent years, to health professionals in general, this branch of philosophy is best known to them under a sub-discipline called medical ethics or bioethics.[5]

Epistemology in lay terminology is theory of knowledge. In other words, we all want to know when we are justified in claiming that we know something to be the case (such as the earth is round, not flat), that in certain contexts, we only believe something to be the case but that, nevertheless, we are justified in holding them, as opposed to those beliefs for which there appears to be no or no good justification. To put things in another way, we are keen to distinguish knowledge from mere belief, to distinguish justified/warranted beliefs from unjustified/ unwarranted ones,[6] to ascertain what counts as relevant/adequate evidence for our beliefs, and so on. While, in general, we are prepared to admit that some of our beliefs may be justified (that is, true) and others unjustified (therefore false), we appear to entertain no doubt about what we claim to know – so then, is knowledge certain, indubitable, non-revisable while belief is probable and/or revisable? Is mathematical knowledge (that, for instance, 2 + 2 = 4) the same as astronomical knowledge (that, for instance, the sun is stationary and at the centre of the planetary system while the earth moves around it), or as knowledge in physics (such as Newton's three laws of motion), and so on?

We have seen above that we make assertions all the time and that necessarily we must try to separate, so to speak, the epistemological sheep from the goats – we must learn to distinguish truths from falsehoods in our daily existence. However, not all of us could be interested in the more formal and abstract notions of Truth and Falsity – this more abstruse preoccupation is best left to experts, the logicians who, unlike us mere mortals, are not interested in the substance or content of assertions made but merely in the formal relationships between what they call propositions (which they designate by symbols such as p or q) without bothering with what the ps and the qs stand for or refer to. For logicians, the key issue is: given a proposition called p and another called q, if p implies q, if q is true, then does it follow that p is true? Given two propositions, p and q, if p implies q, if p is true, then does it follow that q is true? The average reader could immediately appreciate that such formal abstract reasoning would be considered immensely boring by many lay people.

Formal logic may not be everyone's cup of tea but at least it appears slightly easier to grasp than the next big branch of philosophy, namely, metaphysics. The term is made up of two Greek-derived words, "meta" and "physics". We all claim to know what physics is about, but what could "meta" mean? It means "after", but it has also come to mean "above". Hence "metaphysics", coined initially by Aristotle, literally meaning "after physics", has evolved in modern philosophical discourse to refer to that branch of philosophy which deals with matters beyond physics, that is to say, things which, by implication, are neither observable nor measurable. However, it is not obvious that there is such a domain which philosophers could intelligibly investigate and pronounce upon. One way of coming to grips with what the subject is about is to look at what physics itself studies. It studies the behaviour of physical objects which can be as large as planets or as small as atoms (at least classical Newtonian physics does so). Physical objects are material things. In other words, the metaphysician is interested in studying the notion of material objects or more simply of matter. The universe is made of matter; but does matter alone exist? Maybe minds also exist, apart from matter. We, humans, are obviously made of matter but we also seem to possess something commonly call mind. But what is mind, and what is the relationship between mind and matter? If humans have minds, do some animals, namely, the higher ones, also have minds? Material objects are said to possess certain basic characteristics or attributes, such as shape (in general, material objects have shapes), extension (material objects occupy certain portions of space), motion (material objects, in general, move about in relation to one another). Material objects not only occupy space but also certain portions of time – what is space, what is time? These categories mentioned above (matter, mind, space, time, and so on) constitute the domain of metaphysical enquiry; they claim to be part of the fundamental furniture in the universe. Although it is undoubtedly true that some metaphysicians prefer to be minimalist or deflationist (for instance, materialists claim that only matter exists and that any mind-like feature, exhibited by say even human beings, can be reduced to matter), others opt to be inflationist as they admit that many more metaphysical entities exist than in rival accounts of the universe.

Modern philosophy

As already mentioned, the beginnings of modern philosophy may conveniently, though not arbitrarily, be dated to the seventeenth century in Western Europe. What are its major characteristics in terms of the four main branches of philosophy outlined above?

Logic Classical logic operates with two values – true, false.[7] In today's age of computerization and digitalization, these are represented by the two logical gates, 1 and 0. It has three principles regarded as sacrosanct:

1. **Principle of identity**: A is A.
2. **Principle of non-contradiction**: A cannot be both B and not B (at most one is true): $\neg(p \wedge \neg p)$
3. **Principle of excluded middle**: A is either B or not B (at least one is true): $p \_ \neg p$.

Values While the three principles of logic remain by and large unchallenged and impregnable, values, at the opposite pole, have undergone many changes in the last 400 years. In the area of moral values, one may summarize them in terms of the following distinctions:

Objective and subjective: the former strand holds that values are not arbitrarily upheld as they can be grasped or justified in terms of our rational faculty;[8] hence one can work out objectively, indeed, even quantitatively[9] what is the correct answer to a moral problem. The latter maintains that values vary according to the individual person or group/society, time and place; their differences may often be explained in terms of differences in history, sociology, economics, even ecology – as such, there are no universal constants in human values.

Duties and consequences: the so-called deontological view holds that one's duties and obligations are determined independently of the consequences of carrying them out; the teleological view maintains that an action with foreseen overall good consequences over another with less constitutes one's duty. The former upholds the distinction between duties and consequences; the latter collapses duties into consequences.

"Ought" and "is": this is considered to be the crucial problem in modern moral philosophy, namely, that there is a logical gap between "ought" and "is" propositions, which cannot be bridged, as pointed out by Hume. In other words, strictly speaking, one cannot (rationally) argue that because pursuing a certain course of action would bring about overall good consequences, one, therefore, ought to do it.

Epistemology Knowledge about the natural world is based on evidence provided by the five senses. What the senses cannot ascertain (either directly or indirectly) cannot count as knowledge. This kind of epistemology is called empiricism, and the knowledge it yields is empirical knowledge. However, it goes beyond the merely empirical or factual, as it embodies the criterion for what constitutes knowledge.

Metaphysics Metaphysics, as we have seen, is concerned with the ultimate items of existence in the universe. It is also said to be concerned with the question of what constitutes Reality. In the modern (Western) philosophical system (dominant since its rise in the seventeenth century), something counts as real if and only if it counts as knowledge – as nothing counts as knowledge unless it rests on evidence provided by the five senses, it follows that nothing counts as real which is not accessible through the senses. In other words, its epistemology and its metaphysics are intimately entwined, with the former entailing the latter; hence both are empiricist in character.

In the history of modern philosophy, empiricism is also sometimes referred to as positivism, after the term introduced by Auguste Comte (1798–1857) in the first half of the nineteenth century;[10] sometimes the two terms are joined as "positivism-cum-empiricism". Comte claimed that he had discovered a fundamental law which governs the development of the human mind: each branch of knowledge passes successively through three different theoretical states: the theological or fictitious, the metaphysical or abstract and the positivist or scientific. The first and most primitive, the theological, is concerned with the search for first and final causes. The human mind tries to look for the hidden natures of things by posing the question, why things happen in the way they do, and answer it by postulating divine or supernatural beings in Man's own likeness. There is a storm at sea? Why? This is because the gods are angry. In Greek mythology, the cycle of the seasons is explained in terms of the loves and lives of gods and goddesses, of the comings and goings of Persephone from earth to the underworld where Pluto, her husband, lives. Polytheism progresses to monotheism; monotheism postulates God as the first cause, guaranteeing the existence of everything in the world.

The second stage is still concerned to ask the "why" question; but instead of populating the universe with supernatural agencies modelled on human beings, it postulated secular but hidden constructs such as "forces", "powers", "natures" or "essences". Why does water flow downhill? Because it is in its nature to do so. Why does opium send one to sleep? Because of its "dormative virtue." However, such explanations are trivial as they are really verbal or tautologous in character. Opium sends one to sleep because it sends one to sleep; "dormative virtue" is just a high-sounding name to refer to the observable fact that it does put people to sleep.

It is only when a branch of intellectual enquiry reaches the third, positive or scientific stage that one attains genuine knowledge which is empirical knowledge. In other words, the three stages of intellectual development are part of the notion of progress under modernity based on science. Western Europe had long left the first stage behind but until the seventeenth century, it remained shrouded within the second stage, that is, the European medieval worldview. Hence to achieve the ultimate stage of progress, Europeans must liberate themselves from the superstitious approach of medieval philosophy. This is to say, Europe must engineer a revolution in philosophy, constructing a new worldview to replace the old.

Medieval philosophy is often said to be based on what is called Aristotelianism, constructed in the light of the European recovery of ancient Greek philosophy, whose texts were preserved, worked upon and transmitted by the world of Islamic scholars, centred initially in Baghdad (750 CE) and later, Cordoba (after the Muslim conquest of Spain by the early eighth century CE). This Greek/Arabic knowledge was diffused across the Pyrenees to Bologna, Paris, Oxford in the twelfth and thirteenth centuries CE. Western Europe translated these texts into Latin. A key personality in establishing Aristotelianism in Western Christianity was (St) Thomas Aquinas (1225–74) whose greatest contribution lay in reconciling faith and reason by adapting Aristotle's philosophy for the purpose in hand.[11] However, his influence lay not merely within Catholic theology but also provided a philosophical worldview throughout the intellectual domain in Western Christendom.[12] As a result, Aristotelianism[13] reigned in Western Europe for about four centuries until modern philosophy deposed it. The key features of this philosophy appear to combine the themes delineated above in Comte's account of the first and second stages of intellectual development.

It is important here to note that the positivist use of the term "metaphysical" (in Comte's second stage) is meant to be derogatory and abusive and has nothing to do with what Aristotle, and following Aristotle, what this book understands to constitute metaphysics, one of the main branches of philosophy, as outlined above.[14] By and large, for modern philosophy, to indulge in metaphysics is to talk rubbish, even gibberish – it is at best harmless, empty talk, but at worst, it is the enemy of progress.

Medieval philosophy poses as crucial the why question, and invokes first and final as well as formal causes to answer it. Why does opium send one to sleep? Why does an object like a stone thrown down from a height fall faster and faster as it reaches the ground? We have already seen how the question in the first example is answered – opium sends one to sleep because of its dormative virtue or power. In other words, it is in its nature or essence to do so. As for the question in the second example, it is answered by reference to its metaphysics: (i) the ultimate furniture of the world consists of four elements – earth, fire, water, air; (ii) each of these elements has its own natural home, the natural home of fire being the sky (upwards), of objects such as stone the ground (downwards); (iii) that objects, like us humans, have emotions (anthropomorphism). After an absence from home, as we approach our destination, we get more and more excited by the prospect of reunion with our loved ones, and so walk or run faster and faster; in a similar way, the stone falls faster and faster as it, too, is overwhelmed with joy as it reaches its natural home, the ground.

These essences/powers/virtues invoked by scholastic/medieval philosophy are rejected totally by modern philosophy which regards them as "metaphysical" in the abusive sense of the word; modern philosophy must, therefore, reject "why" questions, as it is impossible to answer them without lapsing into mere verbosity and obscurantism. The only legitimate question to pose is the "how" question, as it can be answered ultimately in terms of the evidence provided by the five senses. Empirical evidence could be further strengthened when one can measure objectively and precisely in mathematical terms what one is observing. In the case of the stone thrown to the ground from above, all we can really do is to observe as well as to measure accurately the rate of its fall. Beyond that, we cannot claim to know more, as the "more" turns out to be out of the reach of observation and quantification.

To further illustrate the difference between the old and the new philosophies, we turn briefly to Aristotle and his explanatory framework. How do we understand an object or a phenomenon? Aristotle said it was in terms of four causes – final, formal, material and efficient. To illustrate them clearly, Aristotle used an artefact, such as the statue of a horse. The sculptor was commissioned by his patron to make the statue (final cause); the sculptor chiselled the marble according to a blueprint of the horse, either carried in his own head or drawn on paper (formal cause); the statue was being carved in marble (material cause); the marble was being chiselled by the sculptor (the efficient cause). Modern philosophy ditches the first two, retaining only the material and efficient causes. How does one explain the indentation of a coastline? One explains it in terms of the kind of rock the coastline is made of (material cause) and of the force of the waves/tides which bash the rock (efficient cause). One should also attempt to ascertain accurately the rate of erosion given its material and efficient causes. There is no need to invoke the anger of Neptune (as the final cause) to stir up the waters of the sea in order to lash the rocks, nor some pre-ordained blueprint of the coastline (the formal cause) in Neptune's head to account for its indented shape.

The dropping of the "why" question in terms of the final and formal causes may be said to constitute one of the major elements in the philosophical revolution and worldview of the seventeenth century as well as provide the foundation for modern science and its methodology.

Conclusion

It is crucial to grasp that philosophy and science are inextricably intertwined. It is not possible fully to understand the latter without relating it to the former. Any truly seismic scientific revolution entails not only overthrowing the old science but also the old philosophy which backs it while, at the same time, installing a new philosophy which lays down the parameters within which the new science may be conducted. Hence modern science and modern philosophy necessarily go hand in hand, just as its superseded rival, namely, medieval science and its philosophy – Aristotelianism/ scholasticism – went hand in hand.

2
Modern Philosophy, Modern Science and Its Methodology

This book contends not merely that science and philosophy in general are inextricably linked, and hence that a science is grounded in a particular philosophy, but also that such a philosophical foundation entails a particular scientific methodology. We have already seen the embryo of this latter claim in Chapter 1; in this chapter, we shall elaborate upon it.

Medieval philosophy, as we have seen, considered the "why" question as well as the full suite of Aristotle's causes to be crucial. This in turn implies that its science, too, must be conducted within such a framework, conforming to its explanatory as well as methodological norms.

On the other hand, modern philosophy has tabooed the "why" question, focussing only on the "how" question; it has eliminated two of the four causes, retaining only the material and the efficient. The philosophy of empiricism/positivism, thereby, lays down the explanatory as well as the methodological framework of modern science. Such a philosophy cannot acknowledge an activity as science unless it conforms to its own specific requirements. Hence modern science must be based on what ultimately could be ascertained and tested in terms of observation through the five senses (and through instruments regarded as their extension) – observations constitute the only reality. Reality is Appearance and Appearance is Reality, as there is no other reality (such as the Platonic world of forms) standing behind the world of appearance/ observation. What "appears" to be the case is, indeed, the case.

Galileo, objectivity and mathematization of nature

Of course, science must be capable of eliminating deception which our senses often play on us; what better safeguard can science rely on than on the ideal of Objectivity, when Objectivity is to be achieved through precise, accurate, mathematical measurements in the study of phenomena? To understand this aspect, one cannot but refer to Galileo Galilei (1564–1642), one of the giants of modern science who, at the same time, also contributed to constructing and developing the modern worldview, its science and its methodology. Galileo used mathematics and mathematical measurement as a tool to study nature and to make it yield up its secrets to us. He was struck by the utility of mathematics in the study of physics. This constituted a departure from Plato's and Aristotle's views of the relationship between mathematics and physics. Plato disparaged the world of physical objects, as these are transient, changeable and subject to decay. True knowledge for Plato is knowledge about objects which are unchanging, immutable, and eternal. Pure mathematical ideas seem to qualify for such a status. Hence Plato thought them alone (the forms apart) worth studying. If physical objects do not behave according to mathematical laws, then physics is not worth bothering with, and in any case, as the objects of its study are imperfect and defective, they cannot be expected to conform to the demands of mathematics. Aristotle, on the contrary, inferred from the very abstract character of mathematical procedure, that mathematics could have nothing to offer to physics, as the latter is concerned with the study of matter and its motion, which mathematics precisely ignores.

For Galileo, mathematics enables one to make calculations which could then be tested to see if they fit observation. If they do not, this should not be construed that either calculations are irrelevant (Aristotle) or that observation is not required (Plato). A bad fit could signal that the scientists have left something out of account and that they should go back to redo their homework. For Galileo, observations and measurements yield scientific facts, and if these conflict with existing philosophical beliefs, it is the old philosophy which should give way to the new science grounded in the new philosophy.

Galileo, a giant of modernity, challenged all the major tenets of medieval philosophy and its science and method. For Galileo, scientific knowledge has nothing to do with knowledge uncovered by the use of reason through grappling with the essences of things. Modern epistemology deals not with essences or final and formal causes but with knowledge obtained via the five senses. As essences are grasped through Reason and given by definitions (according to Aristotelianism), Galileo's method of careful calculation and measurement entails their rejection, as these are not amenable to such treatment. He[1] regarded anthropomorphism (that abiotic nature such as a stone can have emotions and desires) as singularly unhelpful as it is a claim beyond the bounds of evidence – all that one can observe and measure is the rate of fall of the stone which enables the observer/scientist to determine the law of acceleration in precise mathematical terms. All else is irrelevant, superfluous and suspect.

The search for universal laws and the unity of method

Science is not so much interested in finding specific occurrences of a phenomenon as in discovering and formulating universal laws of nature[2] or generalisations which obtain without exception. Science holds that order and regularity can be observed to obtain between phenomena in the natural world. It is the task of science to uncover the regularities that do obtain and to systematize knowledge of them – the ideal of science is a pyramidal structure with as few fundamental laws as possible under which as many phenomena as possible may be subsumed, and that the scientificity of a discipline is to be understood in terms of this relationship of subsumption.

In the main, modern philosophy and its science hold that there are only two types of reasoning which can lay any claim to be rational[3] in the pursuit of such knowledge: inductive or deductive. We have already mentioned Bacon as an early champion of the former in Chapter 1; this is later reinforced by the work of John Stuart Mill (1806-73)[4] which we shall have occasion later in the book (Chapter 11) to elaborate through exploring the notions of cause and experiment. Suffice it to mention briefly here that inductive inference is based on the careful extrapolation from as many specific, though varied incidents as possible (cases of B at separate times and separate locations being observed to follow A) to the generalisation that Bs follow As or that As and Bs invariably go together. Armed with such generalisations, the scientist could then predict that the next A would also be followed by B.

The ability to predict what happens in nature is the epistemic goal also of those who rely on deductive logic. As we shall see later in the book, inductive logic is perceived to be flawed, and hence, cannot be relied on by science; instead, science needs a logic which is considered to be its rival, namely, deductive logic, which yields a logically surer form of reasoning to arrive at predicting phenomena.[5] Their tactic consists of putting forward the following schema in deductive logic which one finds in simple syllogistic reasoning of the kind:

All men are mortal (major premise)
Socrates is a man (minor premise)
Therefore, Socrates is mortal (conclusion).

Such a schema when applied in scientific methodology runs as follows:

(All) water freezes at 0 degrees Celsius (generalisation)
This is a bucket of (pure) water (statement of initial conditions)
Therefore this water will freeze at 0 degrees Celsius (prediction).

In modern scientific methodology, the logic of prediction is the same as the logic of explanation – this is sometimes referred to as the "unity of method" thesis. In other words, to explain why the bucket of water freezes at 0 degrees Celsius or to predict that the water in the bucket will freeze when the temperature drops to 0 degrees Celsius, the same schema applies in both contexts.[6]

Unity of science and reductionism

Modern scientific methodology also endorses what is sometimes called the "unity of science" thesis, which proclaims that ultimately, the less basic science can be reduced to the more basic sciences or, indeed, even the most basic. This implies a hierarchy of the sciences; an example of which broadly starts with physics (sub-atomic, then atomic) at the base, followed by chemistry (inorganic, then organic), biology (as molecular or DNA biology), physiology, medicine. This kind of approach embodies what is also called reductionism. Reductionism in its more or less strident forms is deeply entrenched in modern scientific methodology. Take molecular biology after Crick and Watson's unravelling of the double helix structure of the DNA molecule. In the 1950s and the early 1960s, it was not uncommon for molecular biologists to maintain that cells are mere collections of molecules, or "what is true of *E. coli* [a bacterium] is true of the elephant." Admittedly by the 1970s, such a clarion call of reductionism was no longer fashionable, but as shown in the words of a noted historian of that science, the spirit of reductionism remains strong, though somewhat diminished (Allen, 1979, xiii-iv):

> Contemporary biology is characterized by several important factors. One is the firm belief that all biological problems can ultimately be studied on the molecular level. This view does not maintain that studies at other levels of organization, such as that of the cell, the organ, the whole organism, or the population are of no value. In fact, there is a growing awareness among some biologists that it is equally as important to study these higher levels of organization as it is to study the lower, molecular levels. The view that reduction of a complex biological phenomenon to its simpler components (cells or molecules) is a sufficient explanation has become less prevalent among biologists in the early 1970s. Nevertheless, the revolution in molecular biology in 1950s and early 1960s emphasized the importance of understanding the molecular basis of biological phenomena before trying to approach the larger, higher-level interactions.

In yet another area of biology, it remains not merely strong but undiminished. One important and interesting dispute today is the theory of natural selection in the light of theoretical advances in genetics. Richard Dawkins 1976 and 1986 maintains that the unit of selection is the single gene, while those who disagree – Ernst Mayr 1982, Elliott Sober and Richard Lewontin 1982 – hold that selection is not simply at the level of genes or genotypes, but also at the level of the whole organism.

We need to return to the theme of reductionism in the rest of the book, as it remains a prevailing and major part of modern scientific methodology.

Conclusion

Here is a brief summary of the main aspects (as set out in this chapter) of modern science and its methodology, which follow from the philosophical system in which it is embedded:

1. Scientific knowledge is either the only form of genuine knowledge and rationality (stronger thesis) or the highest form of knowledge and rationality (weaker thesis).
2. As such it is concerned with the demarcation between science and pseudo-science or between rationality and irrationality.

3. Above follows from the philosophical system in which it is embedded, whose epistemology and metaphysics are empiricist in character.
4. It admits, at best, two types of inferences, inductive or deductive; either, however, must issue in predictable, hence, testable consequences in terms of observation and measurement.
5. It adheres to symmetry in the logic of prediction and explanation (under the unity of method thesis).
6. It advocates reductionism in its logic of explanation (under the unity of science thesis).
7. Objectivity is the ideal; following Galileo's pioneering efforts, it holds that Objectivity is best achieved via quantification, that is, the mathematicization of natural phenomena.

3
Category *Volte-face*: Organisms for Machines

In the last chapter, we outlined some key features of modern philosophy, and hence also of modern science and its methodology. However, it is left to this chapter to consider what this book regards as the single most important element of the new philosophy/worldview and its implications for the new science and its methodology. It is the philosophy of mechanism which – even today in retrospect – after its first appearance in European thought for at least three hundred years, strikes one as exceedingly bold and mind-boggling. Such innovation, never known before to humankind, necessarily has widespread implications not only for modern science and its methodology but also even in fields beyond.

What is mechanism?

Surprisingly, it is not easy to pin down the meaning of this term, in spite of its familiarity in general discourse as well as in that of the philosophy and history of science. Different scholars have used it to refer to very different matters. For instance, Dijksterhuis in his masterful book (1961, 3) on the subject does not favour the adjectival term "mechanical" as "it smacks too much of automatic in the sense of thoughtless." Nor does he like the noun "mechanism", derived from "mechanistic", on the ground that it is also used "for the internal construction of a machine." He settles for "mechanicism" to designate the system of thought, going under the label "philosophy of mechanism". However, the title of his book speaks of the "mechanization" of the world-picture. In the book, in the main, he discusses the science of motion, covering ancient Greek, classical and modern; that is, the early Greeks, Islam, Christianity and Medievalism, Galileo, Kepler and Newton (amongst others).

His treatment of the subject, however, appears to imply that there is no radical rupture between ancient/classical physics on the one hand and modern physics on the other. His chapter on technology and science (1961, 241–7) does not really touch on the matter in hand, as he merely pointed to the rise of craftsmen as instrument-makers, artist-engineers, painters, sculptors and architects who designed and built canals and locks, fortifications as well as new instruments. This surge of creativity and activity, especially of the *Quattro-* and *Cinquecento* (in Italy), nevertheless, was purely empirical in orientation – involvement with mathematics and science as a result was intimately contingent rather than essentially entwined. In the opinion of this book, such an account is deeply flawed, as the arguments which follow will soon show.

Other scholars tend to take the view that what is key to the philosophy of mechanism is not so much the modern laws of physics as formulated by the giants of the Scientific Revolution from the seventeenth century onwards, but the image of the machine. This presentation does imply a rupture in physics between the ancient and the modern versions of the science itself. More significantly, it maintains that a radical rupture exists between the older and the newer worldviews. This is to say that Western European civilization systematically rejected the organismic view for the machine view of the world. One would dare hazard to say that, up to the seventeenth century in Western Europe, all cultures in human history had, in the main, adhered to the former, in spite of the fact that their accounts in detail would have differed – some, for instance, were supernatural/magical in conception, with the shaman as the chief mediator in upholding the organismically conceived world, while others were entirely materialistic in conception.[1]

Such scholars are, therefore, on the right track but all the same, from the standpoint of this book, they are, nevertheless, wrong in maintaining that the world-is-machine is but a mere metaphor.[2] This chapter will show why this perspective is crucially flawed, as the rupture is more radical than a change in metaphor; it amounts to what this book calls an ontological or category *volte-face*, that the world – indeed, the universe itself and everything in it – is machine. From this basic axiom, the whole of modern science, its methodology and the peculiar characteristics such a kind of science would possess, can be shown to follow from it.

Ontological or category *volte-face*

In the last two chapters, we have referred to epistemology and metaphysics as two main branches of philosophy behind the old as well as the new sciences. The new epistemology and metaphysics which form the philosophical foundation for the latter, as we have seen, are sometimes referred to as empiricism-cum-positivism, or Scientific Naturalism. This account, so far, remains incomplete, as it has left implicit a crucial aspect of its metaphysics, which is best discussed under the more specific term 'ontology'. The relationship between metaphysics and ontology, from the professional philosophical point of view, is a complicated one the details of which, fortunately, need not concern us here. For our limited purpose, we intend to use the latter term to refer to different categories of being; as such, although we have not so far mentioned the term explicitly in Chapter 1, we have already implied it. We know that there are things such as trees, insects, humans. We also know that there are things called numbers, thoughts, grins. Trees may grow tall or may be stumpy; similarly, a human being may grow tall or be short – in other words, there are tall trees and stumpy trees, tall people and short people. Can there be tall numbers/thoughts/grins/ and short numbers/ thoughts or grins, in the way that tress and people are tall or short? We say that some people (or animals) laze in the sun. But does it make sense to say that parallelograms or numbers laze in the sun?

If it does not, then it is because geometrical figures and numbers do not belong to the same ontological category as people and animals such as lizards and sea-lions. Similarly, while it makes sense to say that leaves are green in spring and summer but red or brown in autumn and early winter, it does not appear to make sense to say that Tuesday is red or Sunday is green. (Indeed, the sentences are well-formed from the grammatical, syntactical point of view, but are they, nevertheless, really meaningful?) Colour is just not something one can meaningfully attribute to days of the week, while it is most appropriately attributed to leaves on a tree. One may insist that one is using language metaphorically when one says that Sunday is green because one associates the day with not having to go to the office, with being able to have a long lie-in, being calm and quiet – attributes which one in turn psychologically associates with the colour green rather than the colour red. *Ex hypothesi*, one is not using language literally in such a context; literally understood, it just makes no sense to say that Sunday is green. To hold that it is intelligible to say that Sunday is green or that parallelograms laze in the sun, while insisting that these assertions are meant literally, is to commit what may be called a category mistake.

The above line of argument shows that invoking a metaphor is not the same as invoking an ontological or category change. When the poet uses a metaphor – my love is a red red rose – he does not mean that literally his love is a rose, a red rose, a rose with thorns, a flower, not a female human being. He is taking poetic licence, actually meaning that his beloved is only like a rose in certain limited aspects, that she is beautiful, fragrant, elegant – just as a rose is beautiful, fragrant and elegant – and not that, like a cut flower, he would have to stand her in a vase of chemically treated water so that she would last a few days longer. On the other hand, an ontological change amounts to lifting something which belongs to one category of being to another category of being – in this context, it is to lift things in the world (biotic as well as abiotic) from the naturally-occurring category into the category of artefacts in general and of machines in particular, from the organismic to the mechanistic mode of being. Scholars who portray what is here called an "ontological *volte-face*" as a metaphor (and nearly all of them do) are profoundly misleading, as an ontological *volte-face* is radically different from the mere invocation of a new metaphor.

Related to this flawed view is another. This holds that the ontological make-over is no more than a conceptual change.[3] Concepts are ideas; as language users, many if not all of our ideas eventually find linguistic expression. A concept may some time have to wait for a convenient term to emerge to express it; but often concepts, in both their non-linguistic as well as linguistic forms, appear more or less simultaneously. For example, the term "male chauvinist pig" was not coined till the flourishing of feminism in the 1960s, but the concept in its non-linguistic expression has existed for many decades, if not centuries. Both aspects in the case of the concept that nature (including humanity) is machine emerged in Western Europe, at first gingerly but later stridently, from the seventeenth century onwards. However, whenever a new concept emerges, one must look behind the appearance to find out the reality it stands for – the change from the organismic to the mechanistic mode of being amounts, as we have argued, to an ontological *volte-face*. It is, therefore, not simply a mere conceptual change, or indeed, even a conceptual revolution. It goes beyond language to ontology; in other words, it is an idea or concept grounded in ontology.

People throughout history and in different parts of the world had always considered objects in the world as they found them, to be naturally-occurring things – that is to say, these things existed independently of themselves. Admittedly, many cultures, too, had attributed these naturally-occurring things to be the handiwork of supernatural beings which they called gods or God. But whether divinely created or not, people had regarded the birds in the air, the fish in the river, the plants growing from the soil, the rivers, the mountains, the volcanoes, the glaciers all to be things which occurred regardless of whether they themselves existed or not.

People had also noticed that while rivers and mountains are lifeless things, birds, fish and plants are living things; and these in many cultures were called organisms. Organic life, they would have observed down the ages, goes through different stages in its cycle – birth, growth/development, maturity, decay, death and to be followed by renewal/rebirth. The oak comes from the acorn fallen in the soil; from a sapling, it grows to become a large tree, and when it matures, it produces acorns which in turn may take root and sprout forth new oak trees. The oak can live up to several hundred years but it, too, will eventually die, and decay to become soil. A salmon is not an oak; but as it is an organism, it, too, undergoes the peculiar stages of its own life cycle. This prompts Aristotle in his philosophy to say that each kind of organism has its own *telos*, which governs its own destiny, its own trajectory.[4]

People throughout the ages also realized that while they could not make mountains, oceans, birds or trees, they could make things out of certain other naturally-occurring things they found around them. For instance, they might find a branch fallen from a tree which they might pick up and then use to get at a ripe fruit hanging high up on a tree. They might tie several small tree trunks together with long pieces of bark peeled from other trees, making a raft and use it to cross a river. These were their tools, the first fruit of their technology.[5] These are artefacts.

We have seen in Chapter 2 that Aristotle used an artefact to illustrate his four causes, for the simple reason that in the case of artefacts, these causes are clear and distinct, whereas in the case of a naturally-occurring thing such as an organism, they are intertwined in fact, and could only be disentangled conceptually in the abstract. Aristotle usefully summed this up by talking about the *telos* of the organism. In the case of artefacts, it is obvious that in many cases, only the material cause is naturally given,[6] while the other three can be clearly assigned to human agents – the example of the raft above shows clearly that the small tree trunks as well as the bark peels used in its construction are found in nature (material cause), while the blueprint of the raft (formal cause), the constructors of the raft (efficient cause), the purpose for which the raft is built (final cause) are the result of human activity which, in principle, can be assigned to different human agents in the majority of cases, although they may also be traced to a single human agent in some instances.

These profound differences between organisms on the one hand, and artefacts on the other, may be said to be ontological in character as organisms and artefacts belong to different categories of being. Borrowing from Aristotle, one can say that the naturally-occurring mode of being exemplified by an organism constitutes intrinsic or immanent teleology (as its *telos* is internal to the being itself), while the artefactual mode of being constitutes extrinsic or imposed teleology (as its *telos* comes from without or is external to it). In the case of organisms (which are naturally-occurring beings),[7] their respective *tele* have nothing to do with human purposes, goals, desires – whereas the *tele* of artefacts are, in contrast, nothing but the expression of human purposes, goals, desires. In other words, artefacts are the material embodiment of human intentionality – it is precisely because we humans (efficient cause), want to cross a wide river or go downstream without having to swim or to get wet (final cause) that we construct a raft, according to a certain blueprint (formal cause), using materials such as tree trunks and tree bark (material cause).

The organism, as the naturally-occurring mode of being, has for millennia been regarded as the dominant mode of being in the world. The artefactual mode of being is the ontological foil of the organic mode of being. The latter (in principle) comes into existence, continues to exist and goes out of existence, independently of humans and their activities. The former comes into existence, continues to exist and goes out of existence solely at human behest. In a world without humans or beings with the same kind of consciousness as that of humans, there are no (human) artefacts;[8] the raft as a raft would no longer exist, only tree trunks and tree bark would remain. Artefacts are the products of human intentionality and human ingenuity in executing intentions into material shape and form.

How do we transform the naturally-occurring to become the artefactual? The task falls to technology. However, the history of technology is a very long one, dating back to as early as Palaeolithic time. So it is not a wonder that one may uncover the most simple, "found" technology such as the adze during such a period, and then later, the more sophisticated, craft-based technology ever since humankind left the Palaeolithic for the Neolithic age, until the science-led high technology of more recent times (that is since the 1840s), such as nuclear technology.[9]

So far we have cited statues or rafts as exemplars of artefacts; but these technological inventions did not provide the final impulse to drive the march of progress and modernity in Western Europe. The motor power, so to speak, was provided by another kind of artefact, namely, machines – such as the windmill, the watermill, but later the steam engine.[10] What are machines? In the terminology of the lay person, a machine may be defined as an engineered structure with moving parts which can be used to perform work for us humans (to grind grain, or transport people or goods from one location to another), using a form of energy (whether water, wind, steam, and so on) to do so.[11] A raft is not a machine as it has no moving parts, although it may do work for us, using the river current as its source of energy; whereas a windmill, a steam engine, a motor car are paradigmatically machines, as they have moving parts, using wind, steam or petrol as the energy to perform work for us.

Up to the beginnings of modernity in Western Europe, humankind had not aspired to impose its will outright on the naturally-occurring world in which it found its existence. Admittedly, to live and to survive, we humans must appropriate parts of nature to serve our ends, using whatever technology we are capable of devising. Human consciousness, through the millennia, was shaped, on the whole, by our relatively limited ability through technology to impose our will upon nature. Nor had any thinker articulated the thought of systematically using technology to subdue nature to make her do our bidding. Yet we find that, by the seventeenth century, if not even earlier, in Western Europe, a daring and bold idea seemed to have entered the consciousness of its intellectuals and elites. This radical step consists of what this book calls an ontological or category *volte-face* – the world in which we find our existence is no longer naturally-given but is an artefact, a machine, and in turn, all the things in it (abiotic and biotic) are no longer considered as naturally-given but are machines.

Although the windmill and the watermill were early striking exemplars of machines in Western Europe, it was the mechanical clock[12] – since its appearance from roughly the thirteenth or fourteenth century – which appeared to stir dramatically the consciousness of European elites. However, although it was the mechanical clock which became the dominant exemplar of the machine, what prompted the ontological *volte-face* was probably not any one kind of machine in particular but of machines and other inventions in general. To understand such impact on the European consciousness in late medieval times, one has to remind oneself what Mumford (1946, 108) has said about what he calls the eotechnic period in the history of European technology. It is, on the whole, an era of creative syncretism. Western Europe collected unto itself the technological innovations of other civilizations, adapted and built upon them. To mention just a few – the watermills, already in place in the earlier part of the Christian era, could be traced back to the waterwheel of the Egyptians, who used it to raise water. The windmill had probably come from Persia in the eighth century. Gunpowder, the magnetic needle, paper, and printing came from China, very probably *via* the Arabs, and later, the Mongols. Europe, by 1000 CE, was ready to receive these and other inventions (such as algebra from India, again *via* the Arabs). Glass technology (known as far back as the Egyptians), improved and developed, laying the foundation for the development of astronomy, and of bacteriology by Leeuwenhoek in the mid-seventeenth century.[13]

The slew of machines and other inventions which entered Europe, one following upon the heels of another over a relatively short period of time, managed to create a peculiar impact on European consciousness which was missing in the other cultures from which they originated. This is perhaps because of two related reasons: (a) they came from different cultures – Egyptian, Persian, Arab, Chinese, to mention just a few; (b) these machines and other inventions emerged in their original cultures over a much longer period (in some cases even more than a millennium) and were staggered in the timing of their appearance. In contrast, these machines and inventions came into Europe over a relatively shorter period, and ready-made (according to Mumford's eotechnic period, from 1000 to 1750 CE). In particular, the technological inventions from China were disseminated via the Mongol empire which constituted a unified corridor of communication from the Pacific Ocean to the Danube during parts of the thirteenth and fourteenth centuries.[14] These could have played an important role in the European construction of the modern worldview. One scholar (Bala 2006, 55) has recently written:

> the technological discoveries arose in Chinese culture slowly over an extended period of time, so that each invention came to be absorbed and integrated into its wider organic perspective before making way for others. By contrast these mechanical discoveries arrived in Europe over a vey short period of time and impressed the European consciousness sufficiently to trigger a transformation in thinking that reconstructed the universe in the image of a machine.

As already mentioned, the arrival of the mechanical clock in particular appeared to have triggered off the full flowering of the ontological *volte-face* from the organismic mode to the mechanical mode of being. However, as Richard Westfall[15] points out, this transformation cannot be traced to the efforts of any single thinker, but appeared to be a more or less spontaneous movement in reaction to the then extant worldview. All the same, some such as Hobbes (1588–1679)[16], Gassendi

(1592–1655) and Descartes (1596–1650) were more aware and hence, more systematic than others such as Galileo and Kepler.[17]

To understand more fully the new worldview, one would have to say something about Cartesian dualism in terms of mind and body as two different substances. Descartes struggled to reconcile two things which at first sight seemed irreconcilable – the uniqueness of human beings on the one hand and their commonality with the (higher) animals on the other. Humans, like animals, have bodies but they also have minds (variously called Reason or Soul) which distinguish them from animals. Dualism permitted Descartes to cope with this conundrum. All bodies are forms of matter; matter is brute, inert, subject only to the laws of motion understood by the new physics. Under the old dispensation, mind/spirit/Reason/Soul resided in matter. Hence, eliminating these makes matter *par excellence* the object of scientific study. That is why, for Descartes, as animals are only bodies with no Soul they are mere automata.[18] For Descartes, then, humans are unique in two related ways: (a) they possess not only body but also Soul/Reason/Mind; (b) mind and body, nevertheless, are two very different substances, with mind being superior to body. In this way, not only did Descartes succeed in rendering human bodies as matter to be appropriate objects of scientific investigation, but also satisfied the theological requirement that humans uniquely have Souls which have to be saved. As souls have nothing to do with matter, they are beyond the domain of science and its methodology. In this way, Descartes paved the way for science and theology to co-exist, if not always peacefully.

Once this crucial conundrum has been solved by the Cartesian approach, modern science can safely study all bodies, including the human body, as machines. That is why, unsurprisingly, modern medicine was one of the first sciences to be established. As the details of this new science form the central theme of this book, these would be left to later chapters for spelling out. For the moment, it suffices to look at some leading figures on the subject of the ontological transformation of the human organism into a machine.

Let us start with Leonardo da Vinci (1452–1519), a giant of the Renaissance (the period preceding the Age of Modernity) but whose contribution to the category *volte-face* is, nevertheless, impressive. This has been summed up (Veltman 2011) as follows:

> Leonardo wants to understand the principles of motion in the human body. He sees the body as a mechanical device, effectively subject to the laws of mechanics. Indeed his study of the four powers

of nature, namely, weight, force, percussion and movement, is intimately connected with his anatomical studies and he plans to use
his book on mechanical principles and the four powers as an introduction to his studies of the human body. Traditional organic metaphors are thus replaced by mechanical metaphors. But much more
is involved. Models can now stimulate the body, its parts and its
functions. The spine can be compared with the mast of a ship; the
actions of the shoulders and arms, indeed all the basic human movements, compared with weights and balances. These are evocative
images on the one hand. On the other hand they introduce into the
study of the human body a new level of objectivity, distance, even
coldness... Leonardo's exploration of the human body in terms of
mechanical principles leads him to discover a clinical dimension,
where objective medical treatment is separated clearly from the subjective world of human feelings, emotions and passions.

Another was Santorio Santorio (1561–1636), a physician who was of
aristocratic Venetian stock. He died in Venice, though born in what
used to be Yugoslavia, where his father was sent on official duty. Instead
of considering the body and its functions in Aristotelian (and Galenic)
terms, he championed the view that the body was like a clock with
its interlocking parts, the workings of which were determined by their
shapes and positions. In other words, he held that the fundamental
properties of the body and its functions were mathematical ones, such
as number, position and form.

We next turn to a long quotation from Descartes (*Treatise on Man*,
posthumously published first in Latin, in 1662, in Leyden, though written during 1629–1633 in French, because of fear arising from the inqui-
sition of Galileo), as befits his importance as a leading philosopher and
scientist of the Age of Modernity (ed. Ariew 2000, 41–3):

> I assume the body is nothing else than a statue or machine, which
> God forms expressly to make it as much as possible like us, so that
> not only does he give it externally the color and shape of all our
> members, but also he puts within it all the parts necessary to make it
> walk, eat, breathe, and ultimately imitate all those of our functions
> that may be imagined to proceed from matter and to depend only on
> the arrangement of organs.
>
> We see clocks, artificial fountains, mills, and other similar
> machines, which although they are made only by men, are not with-
> out the power of moving themselves in many different ways. And it

seems to me that I could imagine many different kinds of motions in the machine I am assuming to be made by the hands of God, and I could not attribute it so much artistry that you have no reason to think there could not be more. ...

I should like you to consider next all the functions I have attributed to this machine – such as the digestion of food, the beating of the heart and arteries, the nourishment and growth of the members, respiration, walking, and sleeping; [...] that they imitate in the most perfect manner possible those of a real man. I should like you to consider that all these functions follow naturally in this machine simply from the arrangement of its organs, no more or less than the movements of a clock or other automaton follow from that of its counterweights and wheels, so that it is not all necessary for their explanation to conceive in it any other soul, vegetative or sensitive, or any other principle of motion and life other than its blood and its spirits, set in motion by the heat of the fire that burns continually in its heart, and which is of a nature no different from all fires in inanimate bodies.

However, the most famous or infamous essay on the subject was written by Julian Offray de la Mettrie (1709–1751), published in 1748 in Holland, entitled *L'homme machine* (published in English as *Man A Machine*, two years later). Upon its appearance, the book was publicly burned and La Mettrie was forced to flee to Berlin where he stayed until his death in 1751, under the protection of Frederick the Great. He[19] wrote:

The human body is a machine which winds its own springs. It is the living image of perpetual movement. ... Let us now go into some detail concerning these springs of the human machine. All the vital, animal, natural, and automatic motions are carried on by their action. Is it not in a purely mechanical way that the body shrinks back when it is struck with terror at the sight of an unforeseen precipice, that the eyelids are lowered at the menace of a blow, as some have remarked, and that the pupil contracts in broad daylight to save the retina, and dilates to see objects in darkness ... ?

The human body is a watch, a large watch constructed with such skill and ingenuity, that if the wheel which marks the second happens to stop, the minute wheel turns and keeps on going its round, and in the same way the quarter-hour wheel, and all the others go on running when the first wheels have stopped because rusty or for any reason out of order.

To be a machine, to feel, to think, to know how to distinguish good from bad, as well as blue from yellow, in a word, to be born with an intelligence and a sure moral instinct, and to be but an animal, are therefore characters which are no more contradictory, than to be an ape or a parrot and to be able to give oneself pleasure.

Let us then conclude boldly that man is a machine, and that in the whole universe there is but a single substance differently modified.

The impact of the ontological make-over from the naturally-occurring to the artefactual mode of being was widespread, providing not only the metaphysical basis for modern science and its methodology but also penetrating the domain of natural theology.[20] Take as one example Hume's straw man, Cleanthes, who was set up as exponent of the very view that he, Hume (in 1799), intended to demolish – the words put into Cleanthes's mouth are as follows (1998, 15):

Look round the world; contemplate the whole and every part of it: You will find it to be nothing but one great machine, subdivided into an infinite number of lesser machines, which again admit of subdivisions to a degree beyond what human senses and faculties can trace and explain. All these various machines, and even their most minute parts, are adjusted to each other with an accuracy which ravishes into admiration all men who have ever contemplated them. The curious adapting of means to ends, throughout all nature, resembles exactly, though it much exceeds, the productions of human contrivance; of human design, thought, wisdom, and intelligence. Since, therefore, the effects resemble each other, we are led to infer, by all the rules of analogy, that the causes also resemble; and that the Author of Nature is somewhat similar to the mind of man, though possessed of much larger faculties, proportioned to the grandeur of the work which he has executed. By this argument *a posteriori*, and by this argument alone, do we prove at once the existence of a Deity, and his similarity to human mind and intelligence.

The point of citing this famous passage is, from the standpoint of this book, not about the so-called design argument for the existence of God, but as illustration of how the category *volte-face* had entered even theological discourse – the world which the Almighty had created was nothing but one vast machine, made up in turn of a series of smaller machines.

William Paley's later, even more well-known contribution in his book *Natural Theology* (1802), demonstrates a similar point, but with this difference – in Paley's case, he actually wanted to infer from the watch (with its intricately related mechanical parts) to the existence of the watchmaker in the same way as the eye (with its intricately related mechanical parts) would lead one to infer that the socket (the organism of which the eye is but a part) must have a maker, namely, a divine one.[21]

> In crossing a heath, suppose I pitched my foot against a *stone* and were asked how the stone came to be there, I might possibly answer that for anything I knew to the contrary it had lain there forever; nor would it, perhaps, be very easy to show the absurdity of this answer. But suppose I had found a *watch* upon the ground, and it should be inquired how the watch happened to be in that place. I should hardly think of the answer which I had before given, that for anything I knew the watch might have always been there. Yet why should not this answer serve for the watch as well as for the stone? Why is it not as admissible in the second case as in the first? For this reason, and for no other, namely, that when we come to inspect the watch, we perceive—what we could not discover in the stone—that its several parts are framed and put together for a purpose... [The requisite] mechanism being observed... the inference we think is inevitable, that the watch must have had a maker. Every observation which was made in our first chapter concerning the watch may be repeated with strict propriety concerning the eye, concerning animals, concerning plants, concerning, indeed, all the organized parts of the works of nature. ... [T]he *eye* ... would be alone sufficient to support the conclusion which we draw from it, as to the necessity of an intelligent Creator. http://www.ucmp.berkeley.edu/history/paley.html

Even today such discourse is not without resonance – witness the talk about intelligent design in the debate which also involves Richard Dawkins's defense of Darwinian natural selection in his book, *The Blind Watchmaker* (1986). Dawkins would have no objection in principle to conceiving the world, as well as the various entities in it, to be very complicated machines, except that their complexity as machines requires no divine creator. Indeed, in the view of two other recent theorists of biology (Maturana and Varela), organisms are "autopoietic machines". In other words, organisms are self-organizing machines, but machines nonetheless.[22]

Furthermore, such discourse has subverted even the very expression of a point of view whose explicit end is to combat it. Take the following example:

> Douglas explained that his journey to Madagascar had lit a fire within him that would not go out. In the company of a zoologist called Mark Carwardine, he had found and photographed the elusive lemur known as the aye-aye, an experience, together with reading Dawkins, that had made him realise that the *technology* that now most excited him was the one that had evolved over millions of years and resulted in him and me and, ultimately, the device that wouldn't stop going "boing". He really wanted to understand this business of life and extinction. He and Mark had hit it off straightaway, and the plan was now to find seven more species like the aye-aye that were in imminent danger of disappearing for ever. [The italic is inserted by the author of this book.]

These are the words of Stephen Fry (2009) who published in June 2009 with Mark Carwardine a sequel to the original book of 1990 by Douglas Adams and Mark Carwardine. The subject matter of these two books is a lament upon the extinction of species and their remit is to raise awareness about the potential threat to the extinction of endangered species. Species in the wild (as opposed to domesticated ones) are *par excellence* the results of natural evolution and selection, and hence, are naturally-occurring phenomena. Yet Fry has lapsed (even without realising it) into characterising the process of natural evolution and its issues as mechanical processes and technological products. How ironic that he appears to be unaware that the naturally-occurring is the ontological foil of the artefactual/technological mode of being. Fry's non-ironic use of the word "technology" in the sentence quoted above is evidence of the clearest kind of the depth to which the ontological/category *volte-face* has penetrated contemporary consciousness.

The category *volte-face* and the ideological goal of the new science

This category transformation went hand in hand with the emergence of the new science and its associated ideological goal, which is, to use science (and later its induced technology[23]) to control nature. We have already mentioned earlier the two epistemological goals of science, that is, to explain and to predict phenomena, using the same logical schema.

Take the last – the possibility of prediction leads to the possibility of control of the phenomenon predicted. Comte was clearly of this opinion.[24] If one can successfully make predictions with the help of laws, then one can take steps to get out of the way of the event predicted, if it is considered to be undesirable (the weak sense of control). Or one could alter or modify the circumstances, so that certain desired results could be brought about and other undesired ones prevented from arising (the strong sense). Astronomical knowledge enables one, for instance, to predict an eclipse of the sun at a certain place and on a certain date. Then one can arrange to be there to observe it, if its observation can be used to further some other task, such as Eddington's expedition in 1919 to test Einstein's theory. Alternatively, if an eclipse of the sun is considered to have undesirable effects – suppose observing one causes cancer of the eye – then one could take appropriate avoiding action.[25] The second possibility allows one to interfere more directly with the workings of nature. According to the laws established about plant growth, a certain degree of warmth, and not merely exposure to light, encourages plant growth. If one wishes to encourage growth, then one ought to put the plants in a light, warm place.

The possibility of control in both the weak and the strong senses provides the link between science and technology. In this way, the new science has always been entwined with utility (for humans), a theme that Bacon had made familiar. To Bacon's voice on this matter, let us add that of an even more powerful advocate, namely, Descartes. In 1637, he wrote (1992, 142–3)[26]:

> as soon as I had acquired some general notions in physics and had noticed, as I began to test them in various particular problems, where they could lead and how much they differ from the principles used up to now, I believed that I could not keep them secret without sinning gravely against the law which obliges us to do all in our power to secure the general welfare of mankind. For they opened my eyes to the possibility of gaining knowledge which would be very useful in life, and of discovering a practical philosophy which might replace the speculative philosophy taught in the schools. Through this philosophy we could know the power and the action of fire, water, air, the stars, the heavens and all the other bodies in our environment, as distinctly as we know the various crafts of our artisans; and we could use this knowledge as the artisans use theirs for all the purposes for which it is appropriate, and thus make ourselves, as it were, the lords and masters of nature.

This is desirable not only for the invention of innumerable devices which would facilitate our enjoyment of the fruits of the earth and all the goods we find there, but also, and most importantly, for the maintenance of health, which is undoubtedly the chief good and the foundation of all the other goods in this life. ... we might free ourselves from innumerable diseases, both of the body and of the mind, and perhaps even from the infirmity of old age, if we had sufficient knowledge of their causes and of all the remedies that nature has provided.

As Descartes, indeed, is acknowledged as one of the giants (both as philosopher and scientist) of modernity, it would not be excessive to cite a similar passage from another part of his writing (*Rules for the Direction of the Mind*, 1628):[27]

I think that I cannot keep secret [the rules for the new science] without committing a sin in connection to the law that commands us to seek the good of mankind. Because the rules obligate me to see that it is possible to acquire types of knowledge that are very useful for life, and that, instead of Scholastic philosophy which is taught in the Schools, we can find a practical philosophy with which we may come to know the power and the operation of fire, water, air, the stars, the heavens and of all the bodies that *'environ us'*, as clearly as we know the various crafts of the artisans and manufacturers; we can then, in the same way, make use of these for all the applications to which they might be adopted, and *thus transform ourselves into masters and proprietors of nature.*

The above bears out without doubt the Baconian dictum that "knowledge is power." As such, it would be fair to conclude that built into the new scientific method and its accompanying philosophy from the seventeenth century onwards is the aspiration to control and manipulate (and in that way to dominate) nature. Descartes, Bacon and other formative thinkers of modernity all unhesitatingly subscribed to the ideological goal of the new science. It does not look as if the ideal of knowledge for its own sake, what Einstein called "the holy curiosity of inquiry" ever existed in its neat purity at the inception of modernity (or at any time, later, for that matter). The philosophical, as well as the ideological, requirements of the new worldview ensure that science as technology, and science as theoretical knowledge, go hand in hand. While humans had used and controlled nature in the past, modern science makes it possible for them, more systematically than ever before, to control (to exploit) nature.

Heidegger and science as theoretical technology

The account of science pursued above is akin in spirit to the deconstruction of science as theoretical technology given by Heidegger (1949/1954). From Heidegger's standpoint, the science and the technology appear to be inextricably linked – the linkage is more than an accidental one. As such, it is more than merely contingent. It is, then, not surprising that Science should eventually spawn successful Technology, even though the project of modern science itself took over two hundred years, since its inception, "to deliver the goods", so to speak.

To quote Mitcham (1994, 52–3):

> For Heidegger what lies behind or beneath modern technology as a revealing that sets up and challenges the world is what he calls *Ge-stell*.
>
> *Ge-stell* names, to use Kantian language, the transcendental pre-condition of modern technology. ..."*Ge-stell*" refers to the gathering together of the setting-up that sets up human beings, that is, challenges them, to reveal reality, by the mode of ordering, as "*Bestand*" or resource. ... "*Ge-stell* refers to the mode of revealing that rules in the essence of modern technology and is not itself anything technological." Not only does *Ge-stell* "set-up" and "challenge" the world... it also sets upon and challenges human beings to set upon and challenge the world. ... "The essence of modern technology starts human beings upon the way of that revealing through which reality everywhere, more or less distinctly, becomes resource."

Michael Zimmerman (1990, 181–2), more or less, also makes the same point:

> Far from being a dispassionate quest for truth, scientific methodology had become the modern version of the power-oriented salvific methodologies developed in the Middle Ages. Hence, Heidegger argued, even though modern science preceded the rise of modern technology by about two hundred years, modern science was already essentially "technological" in character, i.e., oriented toward power. ...Science...seeks not to let the entity show itself in ways appropriate to the entity in question, but instead compels the entity to reveal

those aspects of itself that are consistent with the power aims of scientific culture.

Hans Jonas (1966, 189–90), too, has written in the same vein about Bacon's view of science:

> Theory must be so revised that it yields 'designations and directions for works,' even has 'the invention of arts' for its very end, and thus becomes itself an art of invention. Theory it is nonetheless, as it is discovery and rational account of first causes and universal laws (forms). It thus agrees with classical theory in that it has the nature of things and the totality of nature for its object; but it is such a science of causes and laws, or a science of such causes and laws, as then makes it possible 'to command nature in action.' It makes this possible because from the outset it looks at nature *qua* acting, and achieves knowledge of nature's laws of action by itself engaging nature in action – that is, in experiment, and therefore on terms set by man himself. It yields directions for works because it first catches nature 'at work.'
>
> A science of 'nature at work' is a mechanics, or a dynamics, of nature. For such a science Galileo and Descartes provided the speculative premises and the method of analysis and synthesis. Giving birth to a theory with inherently technological potential, they set on its actual course that fusion of theory and practice which Bacon was dreaming of.

In the light of the above and of the points raised in the earlier section, there is, perhaps, some justification in saying that Modern Science is Theoretical Technology.

Conclusion

This chapter has argued that if one were to single out one feature at the core of the rise of the new philosophy in the seventeenth century which enables the new science and its methodology to emerge, it would be the bold, hitherto unarticulated category or ontological transformation of what is natural in general, and of organisms in particular, to become the artefactual in general, or machines in particular.

Accompanying this is the equally bold assertion that the ideological goal of the new science and its methodology is to control and manipulate nature via its technology, in order to dominate it for the benefit of humans. In these ways, the ontological makeover enables human-kind to impose its intention, purpose and goal upon nature. Nature is no longer considered to possess its own *telos*; as matter, it is brute and inert – such would then be material *par excellence* which humankind can fashion according to its own specifications. Extrinsic/imposed teleology would replace intrinsic/immanent teleology. *Homo faber* may be said to be the new god, restructuring the world to its own image.

However, the new mechanical philosophy of the seventeenth century is often not presented in the way above but only via the following theses:

1. All naturally-occurring phenomena can be understood in terms of two notions, namely, matter and motion.
2. Matter is understood to be brute, inert, as spirit or any other active principle has been expelled from it.
3. All scientific questions are posed within such a framework; it also follows that the answers given to them were also formulated by the language which such a framework endorsed.

These theses, although undisputedly correct, nevertheless have missed out an essential aspect of the new philosophy and the new science, namely, the bold ontological perspective that the whole of nature and everything in it are machines.

4
Machines and Reductionism

Machines are defined as artefacts with moving parts which do work for us; hence, it follows that they have parts. This chapter explores the relationship between this fundamental aspect of machines and the metaphysics of reductionism, which in turn constitutes a crucial part of the foundation of the new science and its methodology.

Engineering and reverse engineering

Machines are paradigmatically engineered products. Typically, they are deliberately structured[1] and manufactured. Take the motor car as an instance of a machine. Let us use Aristotle's four causes to understand the concept: its final cause lies in the human desire to transport ourselves and our goods over land, using not animal but mechanical traction; formal cause (the blueprint) would consist of engineering drawings of such a contraption, including all its parts and how these are related to one another; its material cause would be whatever the material(s) deemed fit for constructing the various parts of the contraption; its efficient cause would be the team of designers, engineers, mechanics, and so on who would ultimately be responsible for executing the blueprint into the reality which is called "the motor car".

The various components are separately manufactured,[2] then assembled according to the blueprint, ensuring that the end product runs to the performance criteria set out in its specifications. This makes reverse engineering feasible.[3] For our purpose, we shall simply define reverse engineering as systematic efforts to take apart a product (any manufactured object, structure[4] or system) in order to understand the principles of how it and its parts are related and function.

Even very young children playing with Lego sets are more or less familiar with these two basic activities, engineering and reverse engineering, although they would clearly be more concerned with the former rather than the latter. However, at the end of the play session, when the set has to be packed away, they would have to take the structured object apart; in this way, they would implicitly be engaged in reverse engineering.

All things manufactured in the history of humankind would overwhelmingly be objects made up of parts, whether some of these parts are movable or not. "Found" technology of the simplest kind, such as an adze or a stick, may not have parts – but we have seen that a raft has parts. When one has "reverse engineered" a raft, one would then be left with say, six tree trunks plus strips of bark. These are the basic elements out of which the raft is constructed.

Our ability to indulge in engineering and reverse engineering gives us the confidence that we are in charge of what we are doing and what we want to achieve. As *homo faber,* we inhabit a world in which we are able to control and manipulate matter in accordance with our intentions and our specifications. However, in the long history of humankind the world of *homo faber* or of artefacts, in comparison with the naturally occurring world in which human existence is embedded, remained relatively limited and not overwhelming. We have seen by the seventeenth century in Western Europe, there had arisen a new spirit which conceived of progress in terms of indefinitely enlarging the sphere of artefacts; the aspiration of this new Faustian *homo faber* was/ is based on its ambition to use the new science and its methodology to generate an infinitely more powerful technology which is science-induced to replace the craft-based technology of preceding centuries. In, this way, science and technology, engineering and reverse engineering are inextricably entwined to further the ideological goal of the control and manipulation of nature, to transform the naturally occurring to become the artefactual.

Admittedly, the project of modern science since its initiation in the seventeenth century has not advanced as much on the engineering front as it has on the reverse engineering front. What could be meant by this claim? Take the science of genetics. Engineering has been somewhat limited in spite of the theoretical advances made in the second genetic revolution of the last century – namely, DNA genetics – which has given us biotechnology. This new technology has enabled us indeed to insert DNA sequences from one organism into another, even when these organisms cross the Animal/Plant Kingdoms barrier. Spectacular though such achievements are, it remains till now only an aspiration to create a new form of life from scratch by assembling particular bits of DNA. So much for the domain of the biotic. In the domain of the abiotic, the same verdict that modern science has achieved more on the reverse engineering than on the engineering front also obtains. However, nanotechnology promises to bridge the gap, as nanotechnology is said to enable us to structure some objects atom by atom. The twenty-first century expects science to deliver on the engineering front at the profound level of atoms: when nanotechnology, biotechnology as well as information technology are likely to combine forces to reach new heights of achievement. As a research programme, modern science may have run into many anomalies but, even after four centuries, it still seems capable of delivering surprises.

Engineering, parts/wholes and reductionism

Engineering demonstrates that one can construct a structured object out of certain basic elements or building blocks; reverse engineering demonstrates that one can deconstruct such a structured object in terms of its basic elements or building blocks. These two processes serve to demystify the relationship between parts and wholes. A structured, manufactured object – be it a building, a windmill, a motor car – is a whole constituted out of its various parts. If the parts were assembled correctly in accordance with the specifications laid out in the blueprint, including performance criteria, then the whole should carry out properly the function for which it has been designed. When the whole fails to function properly, the engineer could take it apart to identify any faulty part(s) or any deficient way(s) in which the parts have been assembled. When a (mechanical) clock fails to work, the clock repairer might find that a key screw is loose, or the spring worn out or not accurately put in place. When these flaws are removed, defective parts replaced, the clock would work again.

The relationship between parts and wholes from an engineering point of view may be said to be a functional one. This is to say: (1) each component has a role to play in the whole; (2) its role can be discharged only if it is connected in a certain way or ways to another component or other components in order to ensure that the whole in turn would carry out the overall function for which it has been designed. In other words, if you have understood these two things, you understand all there is to understand about the structured, manufactured object.

It means, too, that you can not only explain the whole in terms of its parts, but also be able (in principle) to predict the function of the structured, manufactured object by knowing the functions of its various components. This chimes in well with the symmetry thesis about the logic of prediction and explanation in the methodology of the new science.

The ability to explain, as well as to predict, the function of the whole in terms of the functions of its parts in turn leads the new science to endorse the philosophy of reductionism, not simply from the epistemological and methodological points of view of revealing function, but also from the metaphysical standpoint. This is to say that wholes do not exist independently of, or over and above the parts which constitute them. Wholes may appear to be things which exist separately and independently of their parts – but in reality they are not real, only their parts are real. The ultimate items of existence in the universe do not include wholes, as wholes are no more than the sum of their parts. When the engineer unravels a watch by dismantling its parts, the watch (as a whole) would have been dismantled, indeed, so to speak, conjured away – what are left on the tray are simply the parts. Having performed the reverse engineering, the engineers – should they so wish – can begin to reconstitute, non-mysteriously, the whole watch from its dismantled components. In other words, the whole comes into existence and goes out of existence according to the purpose and desire of the engineers. To construct the watch (the whole), they only need to get hold (materially) of the components; they cannot get hold of the whole independently of getting hold of the parts. To understand any material object (including artefacts), recall, too, that the new philosophy and its new science have dispensed with two out of Aristotle's four causes, retaining only the material and efficient causes. To hold that the final and formal causes of an artefact, such as a watch, reside in the watch is to subscribe to a piece of obscurantism. They are assigned instead to some agency outside the watch itself, to certain humans who hold its blueprint (formal cause) in their heads or on a piece of paper, as well as entertain the purpose (final cause) for which it was built. But no amount of dissecting the watch in terms of its parts (material cause) would yield evidence about its formal and final causes, in the way that evidence of its material cause can be ascertained. (This is another way of putting the thesis of extrinsic/imposed teleology.) As for its efficient cause, an observer could readily ascertain non-mysteriously the actual processes of its construction – the watch-maker could be seen making the machine.

However, from an alternative philosophical perspective, wholes may be said to exist over and above the existence of their parts;[5] under the new philosophy, to believe that wholes can thus exist is to hold a "metaphysical" belief, in the abusive sense of that term – that is to say, that such a belief is unintelligible, nonsensical.

The belief may also even be said to commit a category mistake. Consider the following: imagine that you've been introduced to all the couples listed in the last UK census and their children. At the end of the introduction, you then say to your host that you have not been introduced to one couple, namely, the average couple. Your error is tantamount, according to the Oxford philosopher Gilbert Ryle, to committing a category mistake. Ryle, in his famous book *The Concept of Mind* (1949), used as an example a visitor to Oxford University who, at the end of his tour of all the colleges, the laboratories, the libraries (Old and New), the playing fields, and all other facilities ... nevertheless asks to see the University. The visitor has failed to grasp that, having been shown all the different parts of the university, there is nothing more, over and above these, which his host could show him as "the university."[6]

One can discuss this claim about the relationship between wholes and parts linguistically, as well as in terms of logic, in the following way: any statement (or proposition) about a whole can be exhaustively analyzed in terms of a finite series of statements (propositions) about its parts, including the relations between them. This means that the proposition about a whole is but a convenient, short-hand way of referring to the finite series of propositions about its parts. The former is equivalent in meaning and content to the latter. It is analogous to saying that the statement about the average couple in the UK having 1.9 children is but a quick and easy way of saying that Couple A has 4 children, B has 2, C has 1, D has 0 ... N has n. From the census, one counts the number of couples in the country (first column), the number of children each has (second column), adds up the two columns of figures and then divides the total in the second column by the sum in the first in order to arrive at the statistical conclusion about the average couple and its 1.9 children.

Yet another way of making the points above is to say that under the new philosophy and its new science, wholes are held not to possess properties which cannot be completely predicted in terms of knowledge about the properties of their parts. They do not have, what is sometimes called, emergent properties.[7] A simple but often cited example of emergent properties is water, the chemical formula of which is: $H_2 + O = H_2O$. The properties of the water molecule (H_2O) are different from the respective properties of the (two) hydrogen and (one) oxygen atoms; nor could one predict the properties of water from a mere knowledge of the properties of its separate atomic constituents.

In the statistical example just cited, 1.9 children in real terms do not and cannot exist; nor can one literally shake hands with or hug the average couple in the same way as one can shake the hands of Jane and her partner Peter, or give them a joint hug. There is nothing more to be known about the average couple apart from the statistical information that it has 1.9 children. The average couple, as an entity, does not exist and hence is not real and cannot possess properties additional to the ones contained in the given statistical data. Terms which refer to wholes are equivalent to terms such as "the average couple."

Conclusion

Wholes can be reduced to parts without loss or residue, whether speaking from the semantic, logical, methodological or metaphysical standpoint.

1. Metaphysical: wholes are not real and do not exist, while their parts are real and exist; wholes are no more than the sum of their parts.
2. Methodological: to understand wholes and how they function, one only need understand their parts, their functions and the relations between them. Engineering enables one to construct a whole using certain elementary components; reverse engineering enables one to deconstruct a whole into its elementary components.
3. Logical: an exhaustive list of propositions about the parts of wholes entails propositions about wholes; put another way, the two sets of propositions are logically equivalent. It is like saying "2 + 2 = 4"; to assert "2 + 2" but go on to deny that the sum is 4 is to commit a contradiction.
4. Semantic: to replace statement(s) about a whole with an exhaustive list of statements about its parts involves neither loss of meaning nor content.

Reductionism in these forms (to varying degrees) constitutes a key element of the new philosophy, its new science and its methodology. This thesis and the challenge posed to it by its critics are sometimes referred to as the debate between individualism versus holism. The latter claims that wholes are more than the sum of their parts, and have properties which cannot be predicted in terms of or reduced solely to properties of their parts, and so on.[8]

5
Organism a Machine

The title of this chapter is a deliberate echo of La Mettrie's *Man A Machine*. Such resonance is appropriate as La Mettrie and other thinkers, as briefly set out in Chapter 3, had provocatively proclaimed that man, the most developed of all organisms (at least in terms of its brain), is a machine. This proclamation contributed to the philosophical foundations for the emergence of modern medicine. However, it was no more than a proclamation. The hard work in three related aspects remained to be carried out in the intervening four centuries, between the seventeenth century and the present. These three aspects are at the level of (a) technology, (b) basic science, (c) theory and ontology, which together work towards transforming the organism to become an artefact, ultimately, a machine.[1] On all these fronts, the development of the ontological notion of organism as machine reaches its apogee by the last quarter of the twentieth century. This chapter is devoted to outlining some of the highlights in this series of development.

The first agricultural revolution

On the technological front, the long history of the domestication of plants and animals may conveniently be divided into two major stages. These are: (1) the first agricultural revolution, followed by the second stage which in turn may be sub-divided into two further stages – (2a) the first scientific agricultural revolution; (2b) the second scientific agricultural revolution. Only the first stage will be looked at briefly in this section; the second with its two distinct though related revolutions induced by the theoretical developments of the science of genetics will be explored in the two sections which follow.

The first agricultural revolution is usually associated with Neolithic culture from around 8000 to 3500 BCE, although one must not forget that this must itself be based upon a vast canon of knowledge about plants and animals and their properties/behaviour transmitted by their Palaeolithic foraging ancestors.

As the climate warmed, in the glades opened up in the forest, Neolithic women (in the division of labour, it usually fell to the women to be the planters and the breeders) created plots around charred stumps and roots, in which favoured herbaceous annuals could be encouraged to grow. And more importantly, under such carefully protected cultivation in the open, plants hybridized with ease.[2] Hybridization would at first be entirely spontaneous, but perhaps later, some plants might be placed in close proximity with others to encourage the process. Promising hybrids would be carefully selected and nurtured so that eventually they contained more accessible nutriment than their wild forebears. But perhaps, even preceding the cultivation of herbaceous annuals was the cultivation and protection of certain fruit and nut-bearing trees, which could take up to thirty years or more to mature. From this perspective, it might be correct to say that horticulture preceded agriculture and, indeed, rendered it possible, as fruit and nut provided the margin of safety in times of food crises and stresses.

Biologists and ethnologists, on the whole, are of the opinion that the first domestication of animals (preceding the Neolithic period) probably began with the dog, and included such barnyard animals as the pig and the duck. However, in the overall domestication of animals including herd animals, it is very likely that, originally, religious/magical motivations, rather than economic/utilitarian ones, would have played a primary role.[3] The veneration of animals is a well-established practice in human culture throughout its history, and it persists, after all, even to this day in some societies.

From the perspective of this book, the most remarkable thing to emphasize about the Neolithic domestication of plants and animals is that it laid down the basic techniques for improvement, leading to the generation of new varieties.[4] This is to select for breeding plants and animals possessing certain desirable characteristics, with the aim of enhancing and improving those properties. Undoubtedly, over the millenia, the techniques were improved upon and further knowledge, information and skills accumulated. But it would not, perhaps, be too simplistic and distorting to say that artificial selection in breeding, relying on craft-based technology, lasted right up to about a mere 70 years ago.

Indeed, during this very long period of human history, from Neolithic times onwards right to the beginning of the 1930s, which mark the coming of age of agricultural genetics, the peasant/farmer/breeder, as far as plants were concerned, used their craft of selective breeding, by and large, relying on open-pollination, to develop what today are called land-races. These land-races have sustained for millenia (and still do in the non-industrialized societies of today) human populations throughout the world.[5] The scientific hegemony in agriculture is, therefore, no more than several decades old.

The system of agriculture in which the land-races were developed is, of course, designed to improve the quality of the plants or animals as well as to increase yield, characteristics deemed desirable[6] by the farmer/breeder.[7] Apart from intensive selection, numerous other devices such as crop rotation, lying fallow, addition of organic manure, and so on were introduced to sustain and improve productivity. By definition, cultivars, which are the result of intensive selection and breeding over centuries, if not millenia, are human artefacts. Humans in creating them have shown remarkable ability in controlling – understood in the strong form – the procedure and determining the outcome, even though it is true that luck would also have played a propitious part in their long history of development. However, it is also true to say that compared with post-Mendelian cultivars, the control exercised, though very impressive, could not be as thorough and as great as that presented in the light of the understanding about the mechanisms of transmission of inherited characteristics given by Mendelian genetics and its associated sciences, such as cytology. To see precisely what this new understanding is, let us now turn to Mendel's discovery itself.

The first scientific agricultural revolution: classical Mendelian genetics and its technology

Craft-based technology used in the domestication of plants and animals over the millennia is said to be based on haphazard trial and error, is non-systematic and, therefore, today, considered to be pre-scientific, if not unscientific.[8] Their end results may be biotic artefacts, but one could neither accurately predict nor explain them relying on such craft-based technology alone. Any advance had to wait upon the emergence of a scientific theory which could explain the transmission of inherited characteristics from one generation to another.

Such a theory was that of Gregor Mendel (1822–1884), who joined the Augustinian Monastery in Brno in Austria in 1843, which enabled him to pursue his scientific research in earnest from 1856. He completed his series of experiments by 1863, after which his administrative duties as Abbot since 1871 prevented him from devoting time to such a secular activity. He published the results of his work in 1866 in a local journal, although he did take the precaution of sending copies to famous scientific establishments in Europe and the USA as well as to some leading scientists of the time in the field, including Darwin.[9] His research was, however, either dismissed or ignored and it was re-discovered only in 1900 when his achievement was finally acknowledged.

As it turned out, Mendel was lucky in his original choice of peas (*Pisum sativum*) in his study of creating hybrids from its different varieties, as these hybrids appeared to follow what have come to be called Mendel's two laws of inheritance.[10] These are: the law of segregation and the law of independent recombination. They may be illustrated by the diagram below, where two varieties of peas differing in only one character trait are crossed, and the dominant trait is expressed in individuals with either genotype[11] AA or Aa, while the recessive trait is expressed in individuals with the genotype aa.

AA				Parents
Aa				F_1 Hybrids
AA	Aa	Aa	aa	F_2 Hybrids

For instance, Mendel found that two tall plants, when crossed, always gave tall offspring; and so too did two short ones. But when a tall was crossed with a short, the F_1 generation of offspring were all tall. Yet when he crossed these, the F_2 generation of offspring yielded the famous ratio of 3:1, three tall to one short. When he in turn bred this generation of shorts with other short plants, they all bred true. So it struck him that the shortness characteristic, although it did not appear at all in the F_1 generation of hybrids, surfaced unchanged in the next generation.

This led him to assume that each organism had two factors (today they are called alleles) for each inherited characteristic (such as height or colour in flowers), one from each parent. The individual could have inherited two like factors, say, both for tallness, or it could have inherited two unlike factors, say, one for tallness from one parent and one for shortness from the other. When expressed in an individual, it looked as if, sometimes, one factor masked the other. The masking factor was the dominant one and the masked, the recessive.

The achievement of Mendel over his predecessors consists of a combination of the following elements:

1. His methodological approach, it is said, was less that of the naturalist and more that of the physicist of his time. The former tended to be more Baconian, to make as many observations as one could, and then try to detect an underlying pattern to them. The latter first analysed a problem, worked out a solution, and then undertook to test it by means of a suitable experiment. A positive result would confirm (support) the solution/hypothesis; a negative result would refute it. Mendel knew certain salient facts about hybrids, such as uniformity in what, today, we call the F_1 generation, but diversity and reversion in the F_2 generation. What he was looking for was an explanation of these facts. His approach based on precise quantification and experimentation fell within what was later called the *Entwicklungsmechanick* conception (enunciated by the physiologist, Jacques Loeb in 1912) or reductionism, which held that biological phenomena and their complex processes such as development, regeneration, fertilization could all ultimately be explained in atomic and molecular terms.
2. His other related methodological innovation was to bring statistics to bear on the study of heredity. Although Francis Galton (1822–1911), who was Darwin's cousin, also used statistics, he did not combine it with the rigour of experimentation, which Mendel did. Galton's experimental technique left much to be desired. For a start, he did not grow his own plants, handing out his selected and graded seeds to helpful friends who grew them, then harvested the seeds and returned them to Galton. Mendel counted all the offspring from his hybrids. In seven series of experiments dealing with crosses, which varied by one factor alone, he examined 15,347 seeds.[12]
3. Other researchers in the field at the time held that only species created by the Almighty were true, that they had lasted since the beginning of the world, and would continue to last, while garden varieties created by humans were monstrosities, and would have a short existence. Their preoccupation with fixed species and whole entities contrasts with Mendel's analysis in terms of the inheritance of unit-characters.

Mendel was perfectly aware that he was being innovative in his research on plant hybrids. He (1965, 8–9) said that although many had worked on them before, "not one has been carried out to such an extent and in such a way as to make it possible to determine the number of different forms under which the offspring of hybrids appear, or to arrange these forms with certainty according to their separate generations, or definitely to ascertain their statistical relations."

The re-discovery of Mendelian genetics at the turn of the twentieth century promised a lot but in reality took more than three decades before yielding real technological fruits. One scholar, Jack Kloppenburg (1990, 77) says:

> Mendel's work was less a Rosetta Stone providing the key to the mysteries of heredity than a uniquely effective agenda for further research. An understanding of the mechanisms of inheritance was to be a crucial tool for the control of transmitted characters, but before the new science of genetics could really begin to contribute to breeding practice, a host of inconsistencies had to be clarified, interpreted in a Mendelian framework, and unified in a coherent corpus of theory. ... If Mendel was necessary for rapid progress he was not sufficient, and despite the hopes of some there was to be no swift outpouring of markedly superior new plant varieties.

Eventually, by the late 1920s, Mendelian genetics had successfully engendered a new technology, called double cross hybridization.[13] This meant, however, that ordinary farmers would not have the expertise (or the time) for such an undertaking, which included genetic knowledge such as linkage, modifying and multiplying factors, factor interactions as well as sophisticated statistical methods in analysing and interpreting experimental data. From being an art or craft, plant breeding became a scientific technology. Researchers with scientific training attached to agricultural experiment stations (and later to seed companies) took over the task.

This shift to the new division of labour between farmers, on the one hand, and researchers (both in public funded institutes and private seed companies) as plant breeders, on the other, was accompanied by a shift in perspective with regard to the plants themselves. Traditional selection was based on the whole plant but Mendelian thinking was focused on its genetic components. This reductionist change of emphasis from the whole organism to its parts constituted an increase in the level of artefacticity in the new variety that eventually emerged. Up to roughly 1925, when the development of the new technology of double cross hybridization was substantially completed and in place, in the USA, plant breeders saw it as their task to adapt imported plants from foreign parts to native conditions. But adaptation became supplanted by incorporation of foreign or exotic genes into established native varieties so that improved new varieties would emerge. The exotic plant was no longer regarded as a whole to be superior to existing varieties in the country, but was simply to be appraised and selected for certain specific genetic traits deemed to be desirable or superior to established varieties. These traits, then, were transferred and incorporated into the latter.

One can see from the above that the technology induced by the basic science of Mendelian genetics has taken a seriously significant step in ontologically transforming the organism to become a biotic artefact, indeed, a machine, at a radically deeper level of manipulation than was possible under traditional breeding practices. The basic components of the whole organism from the point of view of transmitting characteristics from parents to offspring are now identified in terms of genes, carried on chromosomes.

This deeper level of understanding and intensification of control carries the ontological transformation to yet another new height. This is achieved when the goal of increasing productivity (hence profitability for an industry) involves a change from being labour-intensive to being capital-intensive. Machines would replace humans. From the 1940s onwards, the agricultural industry (soon to be called agribusiness), with the help of the hybridization technology, was ready to become increasingly mechanized. In 1938, machine harvesting covered only 15 per cent of American corn (maize). But by 1945, in Iowa it had jumped from 15 to 70 per cent. Traditional, open-pollinated corn varieties displayed a far greater degree of genetic variability in the field than the double cross hybrid, which exhibited greater phenotypical uniformity, even though this might not be as great as that in the single cross. Individual plants in the open-pollination system ripened at different rates, carried different numbers of ears at different places on the stalk, as well as fell over (called "lodging"). Mechanical pickers could damage over-ripe cobs, miss some lodged plants, not strip properly stalks with unevenly situated ears. Hybrid varieties had then to be developed which ripened at the same rate, bore their ears at a certain specific level and angle, were resistant to lodging – qualities designed to make the plant adapted to suit the mechanical picker. Moreover, these new hybrids had tougher shanks connecting the ear to the stalk, which made manual harvesting more difficult, thus reinforcing the need for machines. In other words, a successful research project for the successful mechanization of harvesting is not so much about designing machines to harvest crops – machines being inherently unsuitable for such a purpose – as designing crops to be harvested by machines. Crop plants are as much intentionally engineered products as machines themselves – the ontological transformation of organisms as naturally-occurring phenomena to become artefacts has reached this intricate level of integration, so that there appears a seamless interface between organism and machine.

The second scientific agricultural revolution: molecular genetics and its technology

Molecular genetics is part of a wider approach called molecular biology. Mendelian genetics and its related sciences, such as cytology (the study of cells), had done a convincing job in putting the study of genes within a materialistic framework by linking seemingly invisible genes with visible chromosomes. However, they never really got beyond that to an account of the chemical nature of the genes. Molecular genetics grew out of dissatisfaction with the limitations of such a programme, attempting to go beyond the still formalistic and abstract Mendelian gene-chromosome framework to tackle how the gene actually worked at the structural, functional and informational levels. This was achieved in 1953 when Francis Crick (1916–2004) and James Watson (1928–) discovered the double-helix structure of the DNA molecule.[14]

Their model was of two DNA strands with backbones made of sugar-phosphate forming the twining twin helixes, held together by bonds between the two pairs of complementary bases. Another way of describing it is to say that it was rather like a winding spiral staircase with the sugar-phosphate backbones as banisters, and the pairs of complementary bases as the steps. As the bases are complementary, the two strands are mirror images of each other. When A occurs on one strand, T occurs on the other opposite to it, and to which it is hydrogen-bonded. A sequence, C A C G, on one strand, stands opposite to the sequence, G T G C, on the other. Each strand, on its own, would be sufficient to produce a complementary strand for pairing. The strand containing C A C G would produce G T G C, while the strand containing G T G C would, in turn, generate C A C G. In this way, the molecule would have reproduced or replicated itself. This model then accounted beautifully for the major genetic, biochemical, and structural characteristics of the hereditary material.

This most recent theoretical advance, involving an understanding of genetics and cell biology at the deeper, molecular level, has engendered a new technology which is commonly called biotechnology. From our point of view, the most significant fact about it is that it is far more radical and more powerful than its preceding Mendelian counterpart; impressive though the latter may be, nevertheless, it does not permit the actual transfer of genetic material from an individual of one species to another individual of a different species. This species barrier, biotechnology is now able to cross, through techniques like rDNA transfer (genetic engineering) and protoplast fusion (cell biology), which can produce transgenic organisms; that is, organisms whose DNA has incorporated genetic material from another species.[15] The barrier crossed is, however, not merely between species but also between species from different Kingdoms (for instance, a plant organism can be made to incorporate a DNA sequence from an animal organism). The power of altering what Marx has called "species being" has arrived since the 1970s. In the words of a botanist from Harvard University, at the 1984 USA National Academy of Sciences Convocation on Genetic Engineering of Plants, that power means (National Research Council, 1984, 12): "We can now operationally have a kind of world gene pool. ... Darwin aside, speciation aside, we can now envision moving any gene, in principle at least, out of any organism and into any organism." This is an acknowledgement, if not an outright declaration, that human-kind, at least in principle, has the ability to transcend or by-pass natural evolution and its processes. A 1975 Nobel Prize winner, the M.I.T. microbiologist, David Baltimore has said outright that humans can now outdo evolution.[16]

Theoretical biology and philosophy

The technological transformation of the biotic as a naturally-occurring being to become an artefactual entity has taken several thousand years, finally to reach this most recent deep level of artefacticity by the last quarter of the twentieth century. The first beginnings of the philosophical basis for this technological transformation occurred roughly four to five hundred years ago. Recently, theoretical biologists-cum-philosophers such as Humberto Maturana (1928–) and Francisco Varela (1946–2001),[17] in the light of molecular genetics/biology and their engendered biotechnology have, in the 1970s, made the most systematic attempt to effect the ontological transformation of naturally-occurring organisms to become mere machines.

A key concept in their writings is autopoiesis, at first sight, not a promising term for their purpose, as it roughly means "self-organising". The word itself comes from the Greek *autos* for "self" and *poiein* for "to produce" or "to bring forth". It is used to characterize that property peculiar to a living entity, existing as an organizational unity and maintaining its identity through self-renewal, self-regeneration and self-generation. (However, although the term "autopoiesis" may be new, the concept itself is shared by a lot of theorists outside biology.[18]) So if organisms are self-organising beings, then this is in keeping with the spirit of Aristotle's view that such beings possess their own telos; indeed if they manifest autopoiesis, would it not appear strange for Maturana and Varela to call them "autopoietic machines"? Is that not a contradiction in terms?

They (1980, 76) assert, in no uncertain terms, that all living systems are machines:

> First, we imply a non-animistic view which it should be unnecessary to discuss any further. Second, we are emphasizing that a living system is defined by its organization and, hence, that it can be explained as any organization is explained, that is, in terms of relations, not of component properties. Finally, we are pointing out from the start the dynamism in living systems and which the word "machine" connotes.

Their exclusion of vitalism is clearly uncontroversial. But can one be equally sure with regard to their second and third theses? They reject the standard definition of machines in terms of their components (material cause) and in terms of the purpose machines fulfil (formal cause), shifting the focus from structure to organization which enables them to give an abstract generalized account, applicable to all machines, irrespective of the type of machines they are and of the components, which enter into their concrete realization as machines. As a result, machines, under their dispensation, need no longer be concrete hardware systems, defined by the nature of their components. In other words, machines are no longer necessarily objects constructed out of abiotic or exbiotic nature. This leaves room for extending the notion of machines to living organisms, to regarding dandelions and yeast as machines.

They next eliminate the formal cause – the purpose – of machines as artefacts. They argue that it is not part of the organizational unity of the machine, merely an invitation to invent them in the first place. In other words, they seem to imply that to understand a car as a machine with a certain organizational unity, there is no need to talk about the function, end or purpose it supposedly serves in inventing and manufacturing it. But might this not strike one as odd, even a "naïve" reaction from their perspective? Surely, a car's organizational unity is controlled by the purpose it is designed to serve, namely, to get from point A to Z by moving itself, the driver and passengers, through a certain portion of space. The machine's organizational unity, one would have thought, would be different – the components would be differently related to one another – if the purpose it is designed to serve were different. Moreover, if the purpose were indeed different, even the components themselves and their properties might have been different. Suppose the machine were not designed as a conveyor of people and their possessions from one place to another but for some other purpose, such as to drill a hole underground. Would the machine have the same components with the same properties and be connected up with one another in the same way as the components which enter into the make-up of a car? Far from the standard account being naive, may it not be Maturana's and Varela's account which appears to be misleading. An artefact's purpose or function is not detachable from its organizational unity; on the contrary, its function or purpose informs the very way in which its components are put together as such a unity. Its construction as well as its existence as an organizational unity cannot properly be grasped without a reference to its purpose or function, which enters inextricably into any adequate account of it, both at the conceptual and explanatory levels.

Furthermore, when a machine, such as a car, breaks down, it is facile to say that its failure to discharge its purpose is not part of the disintegration of its organizational unity. A broken crank shaft means that the car cannot discharge its function, that is, to move at all. The broken crank shaft is the cause of the car's inability to move, and therefore of its failure to carry out its function. The cause and the effect together constitute the disintegration of the car's organizational unity. To restore the latter, the mechanics must be guided by the purpose the machine is designed to carry out, which leads them to identifying the broken crank shaft as the cause of its immobility.

The two theoretical biologists-cum-philosophers have seemingly failed to appreciate that their revised definition of the term "machine" has destroyed what is most distinctive about it. A machine on the standard view is an artefact, designed, constructed by humans to serve a distinctive human end. On their account, a machine is no longer an artefact, designed for a specific purpose. Instead, it is any system with an organizational unity to it but physically expressed. An autopoietic machine is, according to them (1980, 135):

a machine organized (defined as a unity) as a network of processes of production, transformation and destruction of components that produces the components which: (i) through their interactions and transformations regenerate and realize the network of processes (relations) that produced them; and (ii) constitute it as a concrete unity in the space in which they exist by specifying the topological domain of its realization as such a network.

It is therefore clear from the above definition that for them the individual organism is a living system, that all living systems are physical autopoietic machines, and that all physical autopoietic machines are living. In other words, for them (1980, 84): "autopoiesis is a necessary and sufficient condition for a system to be a living one."

Their revised account of organisms as autopoietic machines has similarly expelled teleology just as it has expelled the notion from their account of non-organic, non-living (what they have called allopoietic) machines. However, one must remember that teleology in the two contexts does not mean one and the same thing – in the case of organisms, it refers to what this book has called the thesis of intrinsic/immanent teleology as the telos of an organism is part of its very being, while in the latter, it refers to extrinsic/imposed teleology, as the telos of an artefact (such as a machine) is designed and structured into its very being by humans, and it is this which gives the artefact/machine its organizational unity. In contrast, the four causes in the case of living beings are fused, and may only be separated for the purpose of intellectual analysis. Every living being possesses its own *telos* which informs its identity, and governs its attempts at self-renewal, self-maintenance, its processes of growth, maturity, reproduction and finally decay. In other words, they are paradigmatic autopoietic beings, to use Maturana's and Varela's terminology. However, their term "autopoietic machine" amounts to a self-contradiction, while their other term "allopoietic machine" amounts to a tautology – a machine, standardly understood, *ex hypothesi*, is an entity or a system without an immanent/intrinsic *telos*, as it is an artefact embodying a human purpose designed into it. Such a human purpose is necessarily deliberate and intentional, involving full consciousness. Their revised definition of a machine is conceptually misleading and ontologically flawed.

What then could be the ultimate impulse behind their account? Having first identified organisms correctly as autopoietic beings, they, nevertheless, feel impelled to undermine that very status. That is because the implicit elimination[19] of the essential ontological difference between naturally-occurring organic life and biotic artefacts constitutes the theoretical task of preparing the way for turning living organisms into artefactual living beings *via* technological means.[20] In other words, such an account appears tailor-made for the era of biotechnology.

Biotechnology, operating at a more fundamental level of manipulation than Mendelian whole-organism biogenetic technology, enables humans to fabricate living artefacts. Parts of their genetic components – their material cause – may come from another organism, that is to say, an external source. For instance, a gene from a jelly fish could be inserted into a tomato plant which would enable the latter to resist damage from frost. As a result the form of an organism may also alter – for instance, wingless chickens could be genetically engineered. Their ability to grow and maintain themselves has also been ingeniously commandeered by humans to carry out our ends, not to lead lives dictated by their own tele, but to lead lives designed and engineered by humans. To all intents and purposes, humans are their efficient cause.

The discourse of "autopoietic machines", the programme of biotechnology, underpinned by the reductive sciences of molecular biology and molecular genetics go together as a package at all three levels, namely, ontological, theoretical as well as technological. It is part of the unfolding of the ideological goal of science enunciated roughly 400 years ago at the beginning of modern science, namely, that the theoretical discoveries of basic science be used to engender technologies which would enable humankind to control and manipulate nature to serves its own ends. Organic life has appeared elusive for such a research programme, yet by the last quarter of the twentieth century, this long-anticipated goal has, by and large, been achieved.

Conclusion

Man-is-machine is a deeply embedded theme in modern scientific medicine. The last two chapters have charted in some detail its development from its beginning as programmatic pronouncement at least four centuries ago to the present. The two genetic revolutions of the last century together with their respectively engendered technologies have enabled it to attain maturity. We are at the beginning of the twenty-first century, which is expected to be the century poised to benefit more fully than ever before from the fruits of this theme, now that the three strands – ontological, scientific and technological – have so successfully been inter-woven.[21]

It is obvious that the understanding about the transmission of inherited characteristics provided by these two great revolutions in genetics has tremendous implications for medicine and its therapies; however, these will be raised in the relevant chapters to follow in the rest of the book. The task of this chapter was simply to set out in some detail the transformation in general of organisms to become machines.

Part II
Philosophy and Medicine

6
Human Organism is Machine: MEDICINE

Part II will explore in some detail the implications of the ontological *volte face* for modern medicine, in the light of which it would be clear why modern medicine exhibits the characteristics it does. However, before doing so, let us re-cap what earlier chapters in Part I have argued:

1. The most radical and crucial element in modern Western philosophy is the ontological *volte-face* involved in the view that the whole of nature and everything in it are machines.
2. Machines are human artefacts; artefacts are the material embodiment of human intentionality.
3. The two theses above go hand in hand with the scientific project of understanding nature as Janus-faced; it is not merely to explain natural phenomena and their processes out of "holy curiosity",[1] but also to use science to control and manipulate nature, through the technology induced by theoretical understanding of its subject matter.
4. The ideological goal of modern science is to use its technology to transform the naturally-occurring to become artefacts, so that humankind can bend nature to its will, and in this way to be the master of nature. It follows that nature has no other value than to serve human purposes, be it to advance human material wealth, health or the relatively less materially-oriented project of human self-realization. This is what in environmental philosophy is called anthropocentrism and instrumentalism.[2]
5. In the words of Heidegger, Science is Theoretical Technology.[3]
6. As modern science and its technology are intimately linked, the fundamental concept of modernity is not so much simply the Cartesian *res cogitans* ("thinking substance")[4] as commonly held, but *homo*

faber ("human which makes things"). This modern *homo faber* is a being whose brain is informed by basic scientific discoveries which, in turn, guides the hand to design and construct artefacts/machines in accordance with the theoretical understanding at a deeper and deeper level of the structure of matter, provided by the basic sciences. In this curious way, the modern technologist has come to fulfil the Marxist aspiration of the "whole man" – the hand worker is at the same time a brain worker.

7. It also follows that the paradigmatic scientist may not so much be the theoretical physicist as the engineer – whether the engineer designs and constructs an abiotic artefact such as a space shuttle or a biotic artefact, such as a genetic modified organism (GMO).[5] The engineering team which designs the former must know not only Newtonian physics but space physics, material science, aerodynamics, just to mention a few of a much longer list of the relevant sciences. The biotechnology team which designs the latter must know DNA or molecular genetics, cell biology, and so on.

8. Engineering takes centre stage as Science. This sense of engineering should be written, "ENGINEERING" in order to emphasize its encapsulation of the ontological *volte face* and all the implications which follow from that *volte face* at the heart of the modern project of science. In contrast, the more familiar understanding of engineering (as mentioned in 7 above) is a particular, specific scientific activity, taught in universities and other institutions whose graduates are licensed to build bridges, planes, ships, power stations, and so on. This sense is written as "engineering".

MEDICINE as ENGINEERING, Medicine as engineering

The key chapters in Part I have already shown that medicine as ENGINEERING involves two main theses:

1. The object of scientific investigation is the human body; that body is machine.
2. The ideological goal of such a search for explanation is to keep disease under control, to prolong life, if not to eliminate disease and death altogether.

This sense can, therefore, analogously be written: "MEDICINE". There is the more familiar sense which can be written as "medicine", a specific and particular scientific activity, taught in medical schools and related institutions whose graduates are licensed to practise as doctors who may be general practitioners, specialists in hospitals, research workers in medically-oriented establishments, and so on. As part of their training, medical students in their early (pre-clinical) years are taught basic sciences such as physiology, life sciences in general, human biology, cellular biology (cytology), genetics (medical), molecular biology, biochemistry, anatomy, and so on. This pre-clinical theoretical education is then followed by clinical training which covers areas of patients-and-illnesses such as: history of present and past illnesses; vital signs; examination of various parts of the body (heart, lungs, kidneys, head, brain, eyes, and so on); neurological examination; the upper and lower extremities; in-patient/out-patient clinics (clinical pharmacology, public health and so on).[6] Today, both parts of the curriculum constitute what is sometimes called biomedicine, a term which highlights the application of the principles of the natural sciences especially biology, physiology, biochemistry, biophysics to clinical medicine.[7]

There is another similarity between engineering and medicine; both activities are constrained, on the one hand, by certain specific ethical values and, on the other, by economic ones. For instance, in designing a car, the engineers have to test safety against speed – these are two values which can potentially conflict. Engineers have to work out a compromise between them which is acceptable to society in general – the public do want speed but not such speed as to badly undermine safety. However, in doing so, engineers cannot use humans in an experimental vehicle driven at excessive speed in order to crash it so that they can study the impact such a crash would have on the human body. Instead, they use a dummy, made as close to human specifications as possible.[8] Medicine is subject to an analogous ethical constraint – if a substance is foreseen to have very harmful effects, medical trials should not be carried out even on volunteers.[9]

In engineering, economic costs are part of the specifications for the manufacture of the product. One may be able to achieve greater safety but the cost might correspondingly be too high. Technical efficiency must be tempered by price; a trade-off between costs and benefits must be made. The same is true in medicine – one drug may produce better results than another, but it may be too expensive for general use. A similar trade-off (acceptable to society at large) is made between cost and effectiveness.[10]

Medicine and human-is-machine

According to MEDICINE and medicine, the human organism, in conformity with the reductionist perspective, is no more than the human body which is machine; the goal of studying such a machine is to understand its structure and functions of the parts so that disease/pain can be controlled and life prolonged. From this basic axiom, certain theorems follow:

1. As machines are artefacts made by us humans, they (including the human body machine) are objects which belong to this earthly world, not the supernatural one. All explanations about such phenomena are in natural[11] terms – causes need not resort to supernatural deities or their interventions.

2. Machines malfunction when some of their parts are worn-out, leak, suffer blockages, are badly linked together (because a key screw goes missing), or damaged by an external force/agent (such as someone attacking a car with a heavy hammer, something heavy like a large tree crashing down on a car during a storm). The human-body-machine malfunctions (and the patient goes to see the doctor). Some of its parts may have become worn-out (the heart murmurs, the kidneys no longer efficiently get rid of waste matter) or are badly linked together (the femur slips out of its pelvic socket); the machine may leak (incontinence, excessive bleeding), suffer blockage (the arteries become hardened); the body-machine may be damaged by an external force/agent (a car runs over the road accident victim breaking the legs, a mosquito with the malarial parasite in its blood stream biting and infecting the body-machine).

In abiotic machines, worn-out parts (such as a battery, a screw, a pipe) may be replaced by new ones. The same holds true in the case of the human-body-machine. Some worn-out body parts may be replaced – a heart by-pass or angioplasty may be arranged, livers may be transplanted. Prosthetics may be used to replace missing or worn-out parts. Sometimes, abiotic material may be incorporated into the biotic human machine[12] – plastic and aluminium are often used, and of late even computer chips, such as those which form part of a heart pacer, are inserted into the human-body-machine. Such types of activities fall into the domain of surgery and its related technologies.[13]

It is obvious that engineering necessarily involves technologies. To build the pyramids or Stonehenge, the ancient Egyptian as well as the ancient British engineers would have used whatever technologies were available to them at the time. Technologies evolve and develop from region to region, from time to time. Technologies are also transmitted from one culture to another down the ages. The history of technology[14] shows that for millennia, ever since the adze, technology was either "found" technology or craft-based technology, even though the latter can readily be admitted, all the same, to be capable of great feats of engineering. Science-induced technologies, as we have seen, did not appear on the scene till the modern scientific project was initiated in Western Europe during the seventeenth century, when ENGINEERING emerged. However, between its emergence and that of science-induced technologies in engineering, there is a gap of two and a half centuries or so.[15] Ironically, the steam engine (symbol of the Age of Industrialization) itself gave rise to the basic science of thermodynamics which in turn enabled the steam (and other engines) to become more efficient. While the steam locomotive and the four-stroke internal combustion engine were built by illiterate mechanics, such as George Stevenson, or the science-illiterate such as Nikolaus August Otto (clerk and travelling salesman), the story of technology had taken a different turn in general by the mid 1840s, finally delivering the goods so long promised right at the beginning of the scientific revolution – discoveries in the basic sciences induced sophisticated technologies. For instance, the pioneer-inventors of the airplane, unlike the steam locomotive, were people who studied science – Otto Lilienthal (1891) was a German engineer who specialized in aerodynamics, Samuel Langley (1891) was a physicist and astronomer, the Wright Brothers (Oliver and Wilbur) were in correspondence with such pioneers and with the Smithsonian Institution asking for detailed information about flight experiments.

Since the mid-nineteenth century, ENGINEERING had involved science-induced technologies in major domains, including medicine. The pathways followed by technology in both engineering and medicine are similar except in one respect – in the case of the former, we have seen that the craft-technologies used, until the mid 1840s, were the inventions of people who, by and large, knew no or little physics or chemistry. However, in medicine, the low-level technologies from the seventeenth century to the third quarter of the nineteenth century were invented by educated elites, many of whom themselves were trained and practised as doctors. For instance, Laennec (1819) invented the stethoscope by rolling up a piece of paper, putting one end on the chest of the patient, the other end to his ear. He resorted to such a device because the patient in question happened to be an obese lady, making it difficult for him to listen to her lungs without such a make-shift medium. The stethoscope eventually became accepted not only as an instrument for the purpose of auscultation,[16] but also even as the very icon and symbol of a scientifically trained doctor – it hangs around a doctor's neck long after other diagnostic instruments have overtaken its place, almost like a ritual ornament of status, office but most of all, scientific authority. Another example is the thermometer, an instrument with a history of several centuries. However, the modern user-friendly version was not available until Daniel Gabriel Fahrenheit in 1724 constructed the Fahrenheit scale, with the freezing point of water at the lower end and the boiling point at the higher, and then manufactured a thermometer using such a scale and mercury (a material with a high co-efficient of expansion). After that there were many improvements to the instrument.

By the end of the nineteenth century, medical instrumentation became high-tech. This trend began when the German physicist, William Roentgen, accidentally discovered X-rays in 1895 while experimenting with cathode radiation. He found that the rays could pass through human tissue leaving bones and metal visible. By 1896, clinicians in the United States had already adopted the device, the cathode tube, to produce images of fractures and gun-shot wounds in patients.

Other iconic inventions of this new trend in the twentieth century include the electrocardiograph (ECG) which measures the electrical activity of the heart. Willem Einthoven was the first to come up with a working model in 1901, for which he was awarded the Nobel Prize in medicine in 1924. More recently, in 1973, X-rays were married to information technology to produce the CT scanner. The list of such high-tech inventions is endlessly long, and continues to increase at almost breathless pace. In well equipped hospitals today, worldwide, there are state-of-the-art diagnostic machines which incorporate electronics in some way or other. Theoretical discoveries in the basic sciences today very quickly can induce new technologies – the discoveries in cell biology and of DNA sequences have quickly led to numerous new technologies, such as implantation of the human embryo, pre-implantation genetic diagnosis (PGD, also known as embryo screening).

However, whether medical technology is high or low tech, its aim remains constant, that is, to render diagnosis in medicine more scientific, and therefore, more reliable. In pre-scientific medicine, the doctor relied in the main on the patients' reports about their own experience – whether they felt nauseous, weak in the leg, pain in a certain part of the body, how intense, and so on. These may be called symptoms; from the standpoint of science, such accounts were suspect as they were wholly subjective in nature. The scientific method requires not qualitative but quantitative data so, as scientific medicine emerged, symptoms were replaced by signs. Signs may be defined as those features presented by patients but which would elude the patients themselves. However, doctors with their special training, background and instruments could observe and ascertain these signs in an objective and inter-subjective manner. For instance, the thermometer dispenses not only with the patient's own report of how hot/feverish he feels, but also the doctor's reliance on touching the patient and feeling the temperature, for data which fellow experts, in principle, could agree to be accurate. To count as scientific, data must not only be objective and quantitative in nature, but must be capable of being repeated. A patient might truthfully tell the doctor that s/he feels pain in the abdomen, but only the doctor performing a gastro-intestinal endoscopy can determine for sure whether s/he has a peptic ulcer. Similarly, only scientifically qualified staff can take and interpret an X-ray or a scan of a certain part of the patient's body. Such technologies permit doctors to look "directly" at the patient's innards, superseding touching and palpating. Looking at pictures produced by machines is considered to be scientifically superior – X-rays and ultra sound machines bear this out.

Conclusion

The trajectory of medicine as MEDICINE necessarily is oriented towards high tech as theoretical discoveries in the basic sciences induce more sophisticated and high-powered technologies. It is the "cashing out", at every level of inquiry, of the ontological *volte-face* embedded in the man-is-machine axiom.[17]

7
Biomedicine: Some Sciences

Biomedicine

Since 1985, the medicine examined in this book is increasingly called biomedicine; previously, it had generally been called "allopathic medicine", "Western medicine", "scientific medicine" or "modern medicine". These labels are intended to distinguish it from other medicines which are non-European/Western, non-standard, and by implication, non-scientific. The first label was coined in 1842, ironically by the founder of homeopathy, Samuel Hahnemann, who propounded a type of medicine which the orthodox medical establishment has always regarded to be unscientific, even pseudo-scientific.[1]

The term "biomedicine" was born out of awareness on the part of certain anthropologists of medicine of the existence of other large-scale, systematic medicines such as Indian Ayurvedic and Chinese medicine.[2] These theorists were keen to establish that Western medicine, like all these other medicines, is itself a socio-cultural system;[3] furthermore, like all other medical systems, a key factor is the relationship between medical knowledge (theory) and medical action (therapy). In the case of biomedicine, this relationship takes place within a biologically defined framework where only somatic interventions are intelligible and permitted. This meaning of biomedicine is, in particular, relevant to the standpoint of this book as the fundamental focus of this kind of medicine, following upon the ontological *volte-face* and Cartesian dualism, is the body-is-machine axiom. In the words of Gaines and Davis-Floyd 2003:

> Biomedical representations of reality have been based from its (sic) inception on ... the "principle of separation": the notion that things

are better understood in categories outside their context, divorced from related objects or persons. Biomedical thinking is generally ratiocinative, that is, it progresses logically from phenomenon to phenomenon, presupposing their separateness. Biomedicine separates mind from body, the individual from component parts, the disease into constituent elements, the treatment into measurable segments, the practice of medicine into multiple specialities…

The term "biomedicine" appears eminently more suitable to characterize what this book has previously called "modern (Western) medicine" for the following reasons:

1. It is now a globalized medicine, although its origins were Western European. Modernity itself is also a globalized phenomenon.
2. The traditional/folk medicine it superseded in Western Europe now exists at best only in a shadowy form and can no longer act as a counterfoil to "modern Western medicine".
3. Modern (Western) medicine is, as this book emphasizes, part of modern (Western) science. The latter is founded on the ontological change from the naturally-occurring mode of being to the artefactual/machine mode of being; the former, too, is but an aspect of that profound ontological *volte face*.
4. Modern (Western) science is reductionist; so, too, is modern (Western) medicine. For the former, the most basic or queen of the sciences has been said to be physics (sometimes there are two recipients for this accolade, such as mathematics and physics, or physics and chemistry); for the latter, the honour commonly falls on physiology, biochemistry and since the latter half of the last century, genetics. Today, great hopes are pinned on the science of DNA or molecular genetics for further medical successes.
5. It is undeniably scientific, but its scientific methodology is but an entailment of the modern philosophy in which the science is embedded, as Chapters 1 and 2 in Part I of this book have argued. By acknowledging such a qualification, it is far more accurate and far less dogmatic than the older label "scientific medicine", *tout court*.

For these and other reasons, from this chapter onwards, this book will deploy the term "biomedicine" instead of the ones previously used.

Biomedicine: cleavage between its sciences and its therapies in the long early stage

We have seen that the history of modern science and the history of modern technology do not neatly coincide. The former began its journey in the seventeenth century; the basic science of physics culminated in Newton's three laws of motion. In chemistry, Robert Boyle, in the seventeenth century, set up modern chemistry when he separated it from alchemy; he established what has come to be known as Boyle's Law. Antoine Lavoisier in 1789 put chemistry on an absolutely sure footing by establishing the law of conservation of mass which permits quantification as well as precise predictions to be made regarding chemical outcomes.

As Chapters 4, 5 and 6 have briefly raised *en passant*, science-led or science-generated technology did not occur till the 1840s, when discoveries made with the help of theoretical/basic sciences induced powerful technologies, leaving craft-based technology behind for good. To remind the reader, the history of technology, as we have so far implied, may be divided into two major periods:[4]

Phase I   Relatively autonomous craft-based technology which lasted from probably as early as Neolithic times roughly to the 1840s. Ironically, even during the later part of this long period, the causal arrow ran from the direction of craft-based technology to theoretical science – a spectacular example of this is the discovery of the laws of thermodynamics by Sadi Carnot in 1824 when he tried to improve the efficiency of the steam engine, a machine invented by people who knew no science and were even illiterate as was the case of George Stephenson.

Phase II   Technology is science-led from the 1840s to the present. On the theoretical side, by then, most of the fundamental scientific discoveries had already been made. Regarding electro-magnetism, Faraday, in 1831, found that a conductor cutting the lines of force of a magnet created a difference in potential. This, together with the work done by Volta, Galvani, Oersted, Ohm, Ampere and Henry, provided the theoretical foundation for the conversion and distribution of energy as well as for such significant inventions like the electric cell, the storage cell, the dynamo, the motor, the electric lamp. During the last quarter of the nineteenth century, these were spectacularly translated into industrial terms in the form of the electric power station, the telephone, the radio telegraph. Augmenting these were the phonograph, the moving picture, the steam turbine, the airplane.

That was on the physics front. On the chemistry front, it was the isolation of benzine by Faraday in the 1830s (and later, the use of naphtha) which made the industrial use of rubber possible. Advances in organic chemistry permitted the industrial utilization of coal beyond using it as a direct source of energy. From one ton of coal, one could get 1500 pounds of coke, 111,360 cubic feet of gas, 12 gallons of tar, 25 pounds of ammonium phosphate and four gallons of light oils. From coal tar itself, the chemist produced new medicines, dyes, resins, perfumes. Metallurgy also took revolutionary steps forward; however, aluminum, discovered by Oersted as early as 1825, had to await the arrival of electricity, as the cheap source of energy, before its commercial exploitation became feasible in the last decade of the century. Rare metals were incorporated into the industrial processes – for example, selenium, whose electrical resistance varies inversely with the intensity of light, was used in automatic counting devices and electric door-openers.

From (roughly) 1840, technology no longer induced theoretical advances. Under the new settlement, technology has lost that causal initiative and become, much more so than before, simply the executive arm, so to speak, of pure science.

It took theoretical science nearly two and a half centuries to deliver the long-promised technological goodies, so to speak, when science finally achieved in a spectacular fashion the ideological goal of controlling and manipulating nature, a goal which was designed, as we have seen, into the project of modern science itself.

Roughly the same situation obtained in the historical relationship between the basic medical sciences on the one hand and medical therapy and control on the other. (Control and therapy may be regarded as the analogue of technological deliveries in the discussion above.) The medical sciences did not enable doctors to cure their patients or prevent disease until well into the nineteenth century, as we shall see.

However, before we do that, we need to sketch a brief outline of what doctors relied on by way of treatment and cure, until theory-led preventive measures and cures became available. The traditional therapies used by doctors turned out to be those they had inherited from the days of pre-biomedicine which were part and parcel of a conception of medicine ultimately traced back to Hippocrates (ca 460–370 BCE) and three centuries later to Galen. This was the humoral theory of disease. As such, it was not "solid" medicine; unlike germs, humours were fluids of which the body possessed four.[5] The humoral conception was not finally given up till the late nineteenth century. The suite of therapies associated with it ran even longer, until virtually the mid-twentieth century, for more or less 2000 years. This included emetics (the use of substances which make the patient feel nauseous, wanting to vomit), purgatives (to evacuate the bowels), and to an extent cautery.[6] But the therapy most heavily relied on is blood-letting[7] – this was recommended as late as 1923 by Sir William Osler in that year's edition of his *Principles and Practice of Medicine* (first published in 1892) which for 40 years was the standard textbook of clinical medicine. Three main sub-techniques were used: cupping;[8] leeching,[9] which involved the use of leeches to suck blood from the patient; drawing blood from a vein (hence this technique was called venesection).[10]

One might well wonder why such therapies lasted so late into the modern period, especially when modern science had long undermined their theoretical underpinning, and even in the light of certain important discoveries made, such as the circulation of the blood in 1628, oxygen in 1775, the role of haemoglobin in 1862. Since Harvey, doctors had known that venous and arterial blood are one and the same thing except that the latter contained oxygen and the former did not; yet they continued to believe that letting venous blood cured diseases and, therefore, obligatory, while holding that letting arterial blood would do no good but on the contrary harm the patient and, therefore, must be avoided at all cost. One possible reason is that these theoretical advances in understanding the human machine yielded no innovations in therapy. A doctor could not but offer his patients some form of treatment; though the traditional ones from an outmoded conception of medicine might be inefficacious, nay, even damaging, nevertheless, he would have to rely on them in order to be seen to be doing something.

Biomedicine: some sciences

We shall pick up the story of some science-led therapies later in Chapter 8. For now, we shall turn our attention to look briefly at some of the medical sciences – two early ones are anatomy and physiology and a recent one, namely, DNA genetics. These will be presented through the most important features of the metaphysics and methodology of modern science.

Science, it is said, is repeatable. What is not repeatable cannot aspire to be scientific. To render a result repeatable, one must therefore subject the situation to laboratory or experimental conditions. A laboratory is a deliberately structured and designed environment in which the scientist as researcher is fully in charge; an experiment is an activity conducted within a tightly controlled environment in order to coax a certain outcome from the activity.[11] As objectivity in terms of precision and quantification, is the key methodological requirement, the conditions under which the experiment takes place (including its results) must be capable of being repeated by other scientists at other times and (generally) in other geographical locations; the scientific imprimatur will be bestowed if and only if such requirements are satisfied. Similarly, experimental research in the medical sciences must also be conducted under equally rigorous conditions; this would be the highest norm.[12]

Anatomy

Anatomical dissection of the human body took place in ancient Greece. Galen, too, claimed to do dissection but later scholars were of the opinion that he only dissected animals, not humans. After a lapse of a millennium if not more years, human dissection re-emerged in Western Europe during the sixteenth century with Andreas Vesalius (1514–1564) leading the way, upsetting some of the findings of Galen which he found to be erroneous as they were based on animals. In his book, *The Structure of the Human Body* (this being the first textbook on modern anatomy), Vesalius laid down strict guidelines as to how dissection should be conducted and held that basic knowledge of the human body could only be obtained via dissection. It may seem obvious why Vesalius in the spirit of modern science would lay down such an axiom, but let us go through the exercise to articulate what exactly stands behind it:

1. Actually once we reflect on the matter, his axiom, far from being obvious, is counter-intuitive. Why should dissecting a corpse yield basic knowledge of the human body? Surely, we can come to know a lot about the human body by observing such bodies when they are alive – for instance, we know that the human body can only be sustained by the intake of normal air, nourishing/non-toxic food, clean sweet water, by protection in the form of suitable clothing against inhospitable forms of the weather, avoiding certain dangerous situations which could maim, harm or kill, and so on. Furthermore, we humans, through the ages have disembowelled and killed a good many fellow humans in fights and battles, which had given us ample

opportunities to look at the innards of those slaughtered. We must have observed and known that the human body has a brain in the head, a heart, a pair of lungs, a pair of kidneys, a liver, a stomach, large and small intestines, to mention just the main internal organs. We must also have observed and known that our limbs are jointed, that the spine runs down the back of the body, and so on. Is it then, so obvious, that one must conduct dissection in laboratories to discover "basic knowledge" of the human body?

2. This is because "knowledge" obtained "in the field" so to speak is "anecdotal" and therefore, deemed to be unscientific, as by its very nature the conditions under which such knowledge are achieved are not deliberately and precisely designed and controlled conditions. In other words, "the field" is not the laboratory – knowledge obtained under the latter alone counts as knowledge or "proper/scientific" knowledge.

3. The points made above would also explain why dissecting a dead body can be said to yield "basic knowledge" of the living human body, in spite of the critical difference between the dead as opposed to the living organism.

4. However, the difference between the living and the dead is not considered to be critical in spite of being counter-intuitive, as this perspective follows as a consequence of the ontological *volte-face* and Cartesian dualism. Matter is brute, dead and inert whether the matter to be studied is matter which belongs to a dead or a living organism.

5. It is not a wonder that anatomy became the first science in the history of biomedicine, as a corpse, even more convincingly than a living being, is nothing but a machine (albeit one which is broken and can no longer function). Chapter 4 has already pointed out that when an engineer is confronted by a machine which one might wish to understand better, one must do reverse engineering upon it. That is to say, one must take it apart. Dissecting the human corpse is precisely to take the human machine apart, to get at the different components which make it up.[13]

6. As pointed out earlier in this chapter, no new therapy followed specifically from such anatomical studies.[14] Their real purposes appear to have been two:

(a) The most obvious was to use the demonstration to teach students dissection and anatomy. However, considerable resources were put into this activity. Would such an outlay have been justified if the space constructed were to be used merely for teaching purposes?

It would appear not. A less laudable motive was at work – it was simply to boost the status and standing of the doctor and the university of which he was a member who could claim the label "scientific" for this activity. The intention was to blind the lay public with science and to undermine the esteem of competitors who did not do science in the way exhibited. To achieve such a goal in a dramatic fashion, dissection of the human corpse was presented as theatrical performance, with the elite of the city's population paying to attend such a show "on stage" as the great doctor of anatomy wielded his scalpels. A special theatre was built for the purpose, such as the one, which still stands and can be visited today, in Uppsala (Sweden) – this was constructed in 1663[15] as a kind of theatre-in-the-round. On all sides in steep rungs (the bourgeoisie of the time must have been fitter than today's average tourist as climbing up to "the gods" required some nerve and effort) were the seats and standing places. Right in the middle of the theatre was the demonstration table. Looking down from "the gods" on the spectacle, many near-vertical feet below, must have been quite some experience.

(b) The theatrical demonstration in such an impressive space was not simply to declare one's scientific status and standing but also to affirm the new philosophy behind the new science. Teachers of the new anatomy and the new sciences in general were eager to overturn the old philosophy which had been ensconced in the universities of Europe for a long time. That old philosophy was Aristotelianism. Descartes[16] had used his mighty intellect to combat it; the study of the body via dissection was but another way deployed by the new philosophy to fight its powerful rival. On this point, note that the new anatomy was established in universities throughout Europe where it could take on the old philosophy ensconced there via the new science with the new philosophy standing firmly behind it.

Physiology

Claude Bernard (1813–78) is father to physiology just as Vesalius was to anatomy. He was concurrently appointed to a specially created chair in physiology at the Sorbonne in 1854 and medicine at the Collège de France. His ambition was to go beyond anatomy to physiology, as in his view, anatomy was but an observational science, not an experimental one; to progress, medicine must go beyond the former to the latter.[17]

In 1865, he published *Introduction to the Study of Experimental Medicine*. Louis Pasteur wrote as soon as he had read it: 'Never has anything clearer, more complete, more profound, been written about the difficult art of experiment'.[18] Why should Pasteur lavish such praise on Bernard? Note that this accolade was not only for his numerous contributions to physiological functions but more for the way in which he conducted physiological experiments.[19] Indeed, a historian of science in 1957 proclaimed Bernard as 'one of the greatest of all men of science'.[20]

Their reasons would include the following:

1. He rejected authority whether it be academic or scholastic; like Galileo, he said that when a fact contradicts a prevailing theory, it is the theory, no matter however ancient or prestigious in provenance, that should be discarded or modified.
2. Scientific theories are hypotheses which must be tested to see if they are correct – those borne out by the most facts are the ones which are correct, but such theories are in their very nature never final, and therefore, never to be absolutely adhered to.[21] One could say he had as good as anticipated Karl Popper's philosophy of science:[22] "We can solidly settle our ideas only by trying to destroy our own conclusions by counter-experiments." [23] In other words, scientists must strive to disprove their own theories; when the experiment thus designed contradicts the scientist's own conclusion, the scientist must accept the contradiction provided that the contradiction is proved.
3. Scientists must report all the results of their experiments, not only those which support their hypotheses, while suppressing those which do not, but which on the contrary support those of their rivals.[24] Such scientists are not faithful to the epistemological goal of science as truth (though not as certain and absolute truths) – such practitioners "make poor observations, because they choose among the results of their experiments only what suits their object, neglecting whatever is unrelated to it and carefully setting aside everything which might tend toward the idea they wish to combat." [25]

Bernard is rightly celebrated for his achievements so far mentioned; but from the perspective of this book, they do not go far enough. We must also explore what follows:

1. Anatomy is the first biomedical science for the simple reason that it was the most obvious and easiest domain to study in order to make it yield results in conformity with the ontological *volte face* from the

naturally-occurring/organismic mode to the artefactual/machine mode of being.[26] Bernard recognized that for biomedicine to progress, one must get knowledge about the living, not simply the dead organism, using the new methodology which follows from the new ontology. Now this was the truly difficult task to which Bernard dedicated himself. How could a method which ultimately rests on the assumption that the body is mere matter and that matter is brute and inert be applicable to the study of the living human body? The essence of being alive is constituted by the fact that such a body/organism is capable of performing what today we call physiological functions – it breathes, eats, digests food, absorbs nutrition from the digested material while evacuating what is beyond the body's requirement through the bowels and the urinary system, circulates blood and air, and so on. In other words, Bernard needed to go beyond a crude simplistic account of man-is-machine to a more sophisticated level of operation, to show that no additional principles were required over and above what the new mechanistic, reductionist methodology permitted. He (1957, 35) wrote: "... we must believe in science... we must believe in a complete and necessary relation between things, among the phenomena proper to living beings as well as in all others...."

2. In other words, Bernard had to fight the good fight against the "enemy" whose supporters mounted the barricades to proclaim that the new philosophy and methodology would never be able to deliver convincing results where living organisms were concerned. The enemy was vitalism; it may be simply defined as the view that life cannot be explained in terms of the laws of physics, or of physics-cum-chemistry alone, or of physical and chemical properties, but only in terms of a non-mechanistic force. However, Friedrich Wöhler in 1828 had opened a path in undermining vitalism when he tried to synthesize ammonium cyanate in the laboratory; in doing so, he accidentally converted it into urea, a component of urine. This meant that he had synthesized an organic compound from inorganic ones.[27] All the same, this did not deal an instant death blow against vitalism; on the contrary, great scientists such as Pasteur continued to adhere to it. So Bernard saw that he must seize the opportunity to work harder should the next stage of the fight against vitalism be decisive. This work was to be done not by denouncing vitalism as obsolete philosophy or science, but by demonstrating as convincingly as he could, doing what today is called "good science", to show that the functions performed by the living organism could adequately and thoroughly be explained purely in physical/chemical

terms. He (1957, 60) wrote: "I propose... to prove that the science of vital phenomena must have the same foundations as the science of the phenomena of inorganic bodies, and that there is no difference in this respect between the principles of biological science and those of physico-chemical science." His life-long achievements in delivering such goods testify to his determination and success in establishing physiology as a biomedical science on a sure and secure footing, so that eventually vitalism died a natural death and mechanism/ reductionism triumphed – as the colloquial saying goes, "the proof of the pudding is in the eating".

3. An associated battle Bernard needed to fight is indeterminism, as that possibility also permitted vitalism to rear its head. He, therefore, had to establish that determinism obtained in the domain of physiological phenomena and that within its framework, satisfactory explanations of such phenomena in physical/chemical terms could be found. Determinism and indeterminism are big perennial subjects in philosophy.[28] Fortunately, for our purpose here, we do not need to look into all the complex issues involved, especially of free will. Suffice it to say that Bernard was only interested in a very limited aspect which he called "scientific determinism".[29] He (1957, 60) held that scientific determinism obtains in both the study of inert things as well as of living things and that "... there is absolute determinism in all the sciences, because every phenomenon being necessarily linked with physico-chemical conditions, men of science can alter them to master the phenomenon, i.e., to prevent or to promote its appearing. As to this, there is absolutely not a question in the case of inorganic bodies. I mean to prove that it is the same with living bodies, and that for them also determinism exists." In other words, he had in mind experiments conducted under laboratory conditions; his pre-occupation, then, had less to do with metaphysics than with a specific methodological issue, namely, that of repeatability mentioned already above. This was because he was anxious to set up physiology as an experimental rather than a mere observational science. Experiments could and must be repeated by other scientists in order to ensure their scientific credentials. The set of conditions constituting the designed experiment established in one laboratory when reproduced in another should produce the same results. In that sense, this methodological axiom is a particular interpretation of the notion: same cause, same effect. Bernard might have this to say regarding it, "adhere to it"; if the experiment when repeated does not yield the same result, then this must mean that the conditions have

not been similar and, so far, unidentified factors have been at work; next, identify these and control for them; if further experiments still do not bear out the results, then it should lead the scientist to abandon/modify the hypothesis in question rather than ignore the facts.

The period of Claude Bernard's work to set up physiology as a biomedical science was also symptomatic of a trend which began around the 1830s – increasingly many qualified doctors with the appropriate medical degrees and who were licensed to practise medicine, nevertheless, decided to dedicate themselves full time to cultivate biomedical sciences, severing the traditional relationship between medicine and the art of healing/therapy.[30]

Genetics and its therapies

In spite of Bernard's triumph in putting physiology on a sure scientific footing as a biomedical science, using the methodology entailed by the ontological *volte face* in transforming the natural to become the artefactual/machine mode of being, it remains correct to observe that his scientific achievements, in the short run, had nothing to offer to medicine as succour or therapy. It was not till the twentieth century that some of the biological sciences began to induce technologies following theoretical discoveries in the basic science of genetics.

In Chapter 5, we have already given a brief account of the two revolutions in genetics in the last century and shown how their discoveries reinforce the ontological *volte face* of organism-is-machine. We have also talked a little about the technologies engendered by these theoretical discoveries primarily in the domain of agriculture. We now need to say something about those technologies which have been developed for biomedicine as therapy/succour. One suite of such technologies, such as stem cell research, a key biotechnology, may be conveniently looked at under the label gene therapy.[31]

However, one must distinguish between somatic and germ-line gene therapies. The latter is more radical than the former, and if certain ethical values do not stand in the way, then it would be a far more powerful technology to use to eliminate certain diseases from the gene pool found in the human population. The ethical constraints seem to lie in saying that while it is morally acceptable to remove a gene which causes a deleterious effect or introduce a gene without the deleterious expression into the somatic (body) cell of the patient, it is not morally permissible to insert a gene, even a "good" gene into the germ-cells (egg or sperm) of a person; unlike the first method, the second means that the new gene would be heritable and transmitted to the generations which follow. Society is not comfortable with this idea of so ostentatiously "playing God". The first method, in contrast, simply removes the infliction from the patient without implications for genetic inheritance.

Take the possible treatment of Haemophilia B[32] today. Somatic gene therapy has involved the following stages: medical scientists (a) trace the condition to the lack of the factor IX in the blood of certain individuals which prevents it from clotting, (b) isolate the gene for factor IX, then insert it into a (harmless) virus, (c) remove fibroblasts (cells which form connective tissue under the skin) from the patient; (d) infect these fibroblasts with the treated virus, (e) inject these infected fibroblasts, now carrying the missing gene, back into the patient. An inherited disability is now cured by gene replacement therapy.

Germ-line therapy may itself be divided into two types, one slightly more radical than the other but both undoubtedly more radical than the kind just mentioned above. One technique to eradicate haemophilia from the human gene pool does not involve direct modification of germ cells. It tries to ensure that no sons would be born with the genetic disorder by aborting the foetus once its sex and genetic inheritance have been ascertained. The other technique ensures that no mother who is a haemophilia carrier would give birth to daughters who, in turn, will be carriers.[33] Both these techniques would mean that no males would ever be born with the condition, while gene replacement theory merely means that males born with such an inherited condition would be permanently cured of it. Yet this kind of more radical germ-line intervention has not been universally approved of in all cultures and jurisdictions.[34]

Ever since the completion of the 13-year old Human Genome Project (HGP) in 2003, medical scientists have derived great insights into human genetic material with their implications eventually for therapy and control of diseases,[35] in spite of the fact that many obstacles still remain to be overcome for the full benefits of these and other related theoretical discoveries to finally appear. However, it is fitting here to explore the implications of one of these promised technologies for the methodology behind the project of modern science and in turn of biomedicine.

HGP has discovered single nucleotide polymorphisms (referred to, for short, as SNPs, pronounced "snips").[36] A SNP represents a DNA sequence variation amongst individuals of a population. SNPs can be used to identify individuals who could be vulnerable to diseases such as cancer.[37] In the human population, 99.9 per cent of the genetic material is identical; however, what is relevant to understanding genetic differences between individuals lies in that 0.1 per cent. 80 per cent of SNPs may be found in that minute difference. For instance, individual A may display the sequence GAACCT; individual B, GAGCCT – the polymorphism is A/G. So far, research shows that there are roughly 10–30 million potential SNPs, of which more than 4 million have been identified. However, only partial knowledge about them has been obtained, although it is estimated that 10 million of them are believed not to be inherited independently – sets of SNPs adjacent to one another appear to be transmitted from generation to generation without change in a block pattern.

By studying these block patterns (called haplotypes) associated with disease traits, medical researchers believe that they can eventually produce screening tests to identify individuals susceptible to certain diseases, such as arthritis, Alzheimer's, diabetes, and so on.[38] Furthermore, the new emerging pharmacogenomics is set to engineer and deliver drugs targeting the disease in the individual patient via these SNPs. This entails a revolutionary step forward in the history of biomedical pharmacology. Up to the present, medical drugs are in the main mass-produced and target adults, not children;[39] and even amongst adults, no difference is, in general, made between male or female, young or elderly. Pharmacogenomics promises to produce bespoke drugs, which are expected to be more effective, as well as involving less harmful side effects, and so on.

While such optimistic outcomes are being anticipated, we need right now to draw attention to a problem (crucial in the opinion of this author) regarding these research programmes in relationship to the basic methodology of modern science and biomedicine. We have already referred earlier to the fiercely invoked methodological axiom that experimental results must be repeatable in order to earn the imprimatur of being scientific. Repeatability is nothing more than this: same set of initial conditions, same results. Furthermore, as we shall see later – in Chapter 11, on random controlled trials – to be methodologically impeccable, the researcher must hold steady other relevant conditions, save the one which is under investigation between the experimental and control groups. In other words, one could say that it depends on homogeneity, by and large, existing amongst individuals. For instance, the experiment is designed to determine whether a certain factor, call it X, could make a difference to a certain outcome, call it Y (say the intake of alcohol and liver disease). Researchers would ensure that the experimental group will be matched with the control group for factors known to be relevant, such as sex, age, health condition; the control group consists of non-drinkers (or drinkers whose intake is below a certain determined level per day/week) and the experimental group of drinkers (whose intake per day/week is above that determined level). *Ex hypothesi*, the new pharmacogenomics cannot – at least in the case of humans because of ethical constraints – find experimental and control groups who can be matched in the way just described.[40] It is difficult to find ways round this problem, so that the methodological requirement of repeatability could be met. If it cannot be overcome, would one then have to conclude that pharmacogenomics (for humans) is not scientific? Or would biomedicine have to modify a fundamental methodological canon?

Conclusion

This chapter has attempted:

1. To argue that the term "biomedicine" is the most appropriate term to use with regard to modern (Western in origin but now globalized) medicine.
2. To show (a) why anatomy was the first biomedical science to be established, to be followed by physiology. These two basic sciences have been chosen to illustrate their crucial historical significance in the development of the man-is-machine framework.
3. To intimate that pharmacogenomics, based on the theoretical discoveries of DNA genetics/biology may have the potential to undermine an important methodological requirement of biomedicine as proper science.[41]

8

Biomedicine: Some Technologies

"Deeper" theories, "deeper" technologies and increasing degrees of control

Chapter 6 has explored briefly the notion of instrumentation in biomedicine; its history shows that from a low tech base, it has changed increasingly over the last hundred or so years to a high tech one. Instruments are obviously forms of technologies; this chapter will look at other less obvious forms, such as surgery and pharmacology. This is in keeping with the view expressed earlier that the majority of medical therapies may legitimately be regarded as technological interventions upon the diseased body, as such interventions would involve instruments generated by theoretical discoveries in the basic sciences.[1] However, before doing so, we need to elaborate a bit more on why biomedical technologies are necessarily oriented towards the high tech end of the spectrum.

The point has already been made that the project of modern science (which includes biomedicine) began roughly two centuries before it finally succeeded in rendering its main ideological goal of controlling/manipulating nature bore fruit. We have shown that since the 1840s, theoretical advances in the basic sciences have induced corresponding technologies. Recall, too, that this project, given its philosophical roots and its methodological orientation, is necessarily reductionist. To recap quickly, reductionism explains wholes entirely in terms of their parts, thereby requiring such a science to dig beneath "the surface" to the "reality" behind it. In biomedicine, the explanatory quest is to go beyond the whole body to first, the organs which constitute it; second, to the tissues which constitute the organs; third, to the cells which constitute the tissue; fourth, to sub-cellular levels – namely, of the molecules and atoms which constitute the cells. One can see that investigation of the human body is progressively conducted at a "deeper and deeper" level of the structure of matter.

We have already shown that anatomy may be regarded as the first biomedical science; it investigated through dissection the organs in the human body, and later the correlation between the lesions in the diseased organ and earlier diagnosis of the patient's condition. Physiology looked into the relationship between the organs and their functions. The study of tissues followed – today it is called histology. Cells were then studied – made possible with the invention of the microscope – by Antony van Leeuwenhoek (1632–1723), Robert Hooke[2] (1635–1703) and Marcello Malpighi (1628–1694). In the nineteenth century, Matthias Schleiden (1801–81) and Theodor Schwann (1810–82) established cell theory, today known as cytology, following Rudolf Virchow's dictum in 1855 that "All cells arise from cells." In the twentieth century, biochemistry of the cell developed; today, cell biology integrates both chemical and structural data arising from these approaches.[3]

What could be meant by "deep" or "deeper" theories in science? The term may be understood in at least three ways:

1. A less deep theory is ultimately to be explained in terms of a deeper one. The kinetic theory is explained in terms of the atomic theory, and the latter itself is accounted for by sub-atomic quantum theory. Relatively speaking, the first is less deep than the second, and the second than the third. Similarly, Mendelian genetics is accounted for in terms of molecular genetics.
2. The deeper theory may also then be said to be more comprehensive in scope, explaining a wider range of data, accounting for more variables in their causal contribution to a particular phenomenon.
3. A less deep theory may contain laws about particles and their behaviour at the macro level of existence and observation, while a deeper theory postulates laws about particles and their behaviour at the micro level of existence and observation. Newtonian macro physics may then be said to be less deep than quantum physics.

The modern project of science and technology is built on an ontology of (reductionist) materialism. Ever since its inception, its central aim has been to penetrate the nature and structure of matter. Matter at the macro level of existence is to be broken down analytically into its component parts at the micro level of existence. Hence the atomic theory of matter: all macro objects are made up of molecules which are themselves combinations of atoms; in turn, atoms themselves are to be explained in terms of the sub-atomic theory of matter.

It has been the ideological goal of this project from its very beginning in the seventeenth century to use its theoretical advances to engender powerful technologies to control nature in order to serve human ends. This promise has been made good from the middle to the late nineteenth century onwards. And as its theoretical advances get deeper and deeper into the structure of matter, the theory-induced technologies get more and more powerful.

Take biology as a discipline. In the words of one well-known historian of the subject (Allen 1979, xiii–xiv):

Contemporary biology is characterized by several important factors. One is the firm belief that all biological problems can ultimately be studied on the molecular level. This view does not maintain that studies at other levels of organization, such as that of the cell, the organ, the whole organism, or the population are of no value. In fact, there is a growing awareness among some biologists that it is equally as important to study these higher levels of organization as it is to study the lower, molecular levels. The view that reduction of a complex biological phenomenon to its simpler components (cells or molecules) is a sufficient explanation has become less prevalent among biologists in the early 1970s. Nevertheless, the revolution in molecular biology in 1950s and early 1960s emphasized the importance of understanding the molecular basis of biological phenomena before trying to approach the larger, higher-level interactions.

Biologists, on the whole, since the late 1970s, may, indeed, have resisted strident reductionism of the kind which says that cells are mere collections of molecules, or "what is true of *E. coli* [a bacterium] is true of the elephant", a view prevalent in the 1950s and 1960s. But it remains true that they unanimously agree that molecular biology provides a deeper level of theoretical understanding than classical Mendelian genetics, leading to much more powerful technologies culminating in the creation of human-made life.

Let us go back to the example of haemophilia (briefly examined in Chapter 7) and explore the history of treating the condition to illustrate the relationship between deep theories and deep therapies, so to speak. We can distinguish five stages; the earliest is related to mere observation generating at best what may be called a simple technique of control, whereas the other four are more than simple techniques but are technologies generated by deeper and deeper theoretical under-standing of the condition.[4] The first may at best constitute control in the weak sense of the term; the rest constitute control in the strong sense.

1. This rule of the first stage may be formulated as: if unstoppable bleed-ing is to be avoided, the sufferer of haemophilia ought to avoid get-ting bruised or cut. The scope of this technique in terms of its efficacy is not great as it is useless, should the sufferer, unavoidably, become bruised/cut. There are, unfortunately, many such situations arising in the lifetime of a sufferer. Its efficacy is no more impressive than its analogue in a hurricane context where one could, at best, only advise people to get out of the way of the hurricane, when the signs of its imminence are detected, there being no means of deflecting it or defusing its strength. This minimal degree of control, though based on sound observation, may in some contexts (though certainly not all) be a reflection of the lack of theoretical understanding of the phenomenon in question.

2. The technological rule of the second stage may be formulated thus: to prevent unstoppable bleeding, the sufferer ought to be given blood transfusion containing normal blood of the right type. Undoubtedly, its scope of efficacy is greater than that of the tech-nique of control characterized above, for it can offer succour even after the sufferers unavoidably have bruised or wounded them-selves. But, nevertheless, it would be beside the point should the appropriate type of normal blood be not available for transfusion.[5] The increase in control reflects the theoretical understanding that the condition is caused by an inability of the sufferer's blood to clot, owing to its lack of a certain chemical, and that it is a genetic disability, not a functional one.

3. The technological rule of the third stage may be formulated as fol-lows: to prevent unstoppable bleeding, the sufferer ought to be given the clotting chemical (factor VIII or IX, depending on whether it is a case of Haemophilia A or B). Its scope of efficacy is greater than that of the first technological rule, as it overcomes the scarcity in the supply of normal whole blood, especially when the clotting agent in question can be produced *via* genetically engineered organisms.[6] Also, the clotting agent can be more conveniently introduced into the sufferer's body through injections, rather than the more cum-bersome technology of full blood transfusion itself. This greater degree of control is a reflection of the more detailed theoretical

understanding about the nature of blood in general, and the specific deficiency isolated in the blood of haemophiliacs.

4. The technological rule of the fourth stage may be formulated as follows: to prevent unstoppable bleeding, the sufferer ought to be given gene replacement therapy.[7] Its scope of efficacy is greater than the technological rule at the third stage, as it renders repeated and tiresome injections of the clotting agent throughout the lifetime of the sufferer redundant. And even more tellingly, the sufferer, formerly identified as a haemophiliac, is transformed under such treatment into a non-haemophiliac. His status has spectacularly altered. His genetic disability has been removed once and for all (if the treatment is truly successful). This still greater degree of control reflects yet more advanced theoretical understanding of the nature of heredity *via* classical Mendelian as well as molecular genetics/biology.

5. The technological rule of the fifth stage may be formulated as follows: to prevent unstoppable bleeding in individual males, germ-line therapy ought to be given to the female carriers of the condition. This would yield male genotypes with the gene to produce factor VIII or IX.[8] Its scope of efficacy is in turn greater than that of the technological rule at the fourth stage, for it actually tackles the problem, at an earlier stage, by ensuring that no males would be born haemophiliac in the first place. This ultimate degree of control is a further reflection of knowledge about the nature of haemophilia as a genetic disability based on Mendelian as well as molecular genetics and molecular biology.

One caveat should be entered. The co-relations between the efficacy of technological rules, their corresponding degree of control on the one hand, and theoretical advances in the relevant sciences on the other, as set out above, are not meant to reflect precisely actual historical correlations. They are meant to bring out more clearly the epistemological linkage between technological rules and scientific laws, namely, that laws ground the efficacy of such rules.[9] In so doing, one is also laying bare the philosophical foundations for the ideological goal of biomedicine to control nature in the strongest form possible, to make it serve human ends, those of alleviating pain, eliminating disease and disabilities, and so on.

Surgery

Surgery in various forms has a very long history, with beginnings as early as Neolithic times[10] and in numerous cultures in historic times throughout the world. People whenever and wherever are prone to breaking a bone or two, suffering from traumas such as wounds, especially those inflicted during private fights and wars. Fractures have to be mended, bones re-set, limbs amputated, bleeding stopped, teeth pulled out, and so on. Indeed, the place of surgery in the history of medicine is well expressed by Hippocrates (460–370 BCE): "What cannot be cured with medicaments is cured by the knife, what the knife cannot cure is cured with the searing iron, and whatever this cannot cure must be considered incurable."[11]

As this book is only concerned with the development of medicine in Europe leading to the establishment of biomedicine, this discussion of surgery will also, therefore, be confined within such limits.

In this context, there could be no humbler origin than surgery as from medieval times, the group of people who conducted surgical operations were none other than barbers, who today only cut men's hair, shave their beards or groom their clients in other ways. But for about seven centuries, barbers did more than that; they performed surgical operations as well. As a profession, they were already established by 1094. These practitioners were, therefore, called barber-surgeons.[12] In particular, their expertise was in great demand by the military, as they were needed to look after the wounded; hence they took up residence, in the main, in castles and other fortifications. The separation of surgeons from barbers did not formally take place in England till an Act of Parliament enacted on 2 May 1745.

The journey from lowly craft to science was a slow one. Its early beginnings could be traced back to the sixteenth century when the great barber-surgeon, Ambroise Paré (1510–90), as military surgeon, applied the principles of Vesalius's anatomy to the treatment of war wounds.[13] He pioneered new techniques such as ligature to stop bleeding (although the rate of infection was so high that it was abandoned as a practical measure) as well as accidentally found a more effective substitute (a tincture made from egg yolk, turpentine and oil of roses) for cauterization during amputation. He also developed artificial limbs and some new surgical instruments. He held that the pain in phantom limbs was in the brain, a view which neurology accepts today.

He set out five reasons for surgery which have all come to pass and have not, in general, been surpassed since: "To eliminate that which is superfluous, restore that which has been dislocated, separate that which has been united, join that which has been divided and repair the defects of nature."[14] However, Paré was well ahead of his time. Three major practical issues confronted surgery at the time: pain, bleeding and infection. We have already seen how his technique of ligature had to be dropped as infection was a serious risk; the problem was not overcome until Lister (1827–1912) pioneered antisepsis. A healthy individual can tolerate ten per cent to fifteen per cent of blood loss in the total volume without serious medical difficulties; this is not much, given that about eight per cent to ten per cent is taken normally from a donor today under blood transfusion. However, blood transfusion itself, from human to human, was not successfully undertaken in a major medical operation until 1818.[15] This technique, when refined, would render, by and large, major surgical operations safe.[16] Traditionally, pain was relieved with alcohol, opium, mandrake (if not more gruesome methods such as knocking the patient unconscious with physical force). It was not until 1846 that ether was used in three successful minor operations, and in 1847, chloroform.[17] In other words, until all these anaesthetic and other techniques such as antisepsis and asepsis (based on at least partial if not full theoretical understanding of the relevant areas of research) were in place, surgery could never be a general readily-invoked therapy.[18]

Surgery finally left barbering – and its status as a lowly trade – behind to become a part of biomedicine when, in 1800, the Royal College of Surgeons was established.[19] At the same time, the rise of hospitals since the French Revolution and later the Napoleonic wars[20] also made major surgical operations feasible, as patients with trauma were readily available for surgeons under one roof with the necessary resources to perform such an activity. However, one must bear in mind that the great strides in surgery since the nineteenth century took place within the framework constructed in the light of the ontological *volte face* which put centre-stage the human-is-machine world-view. We have also seen that they were based on the biomedical science of anatomy. This science, we argued earlier, bears out the first fruit of the human-is-machine axiom. Surgery, too, as we shall show, is also eminently susceptible to the human-is-machine ontological perspective.[21]

Surgery is both ENGINEERING and Engineering *par excellence*. Look again at Paré's five reasons for surgery. They are to render the human machine leaner, more efficient, with only fully functioning parts. Hence, (seemingly) redundant parts may be removed. In this spirit, many children in the United States (particularly of middle-class parents in certain states) were routinely subjected to tonsillectomy to prevent inflammation and its associated problems; paediatricians held/hold that this set of lymphoid tissues play no significant role in the immune system.[22] Fractured bone parts must be repaired; amputated limbs must be replaced by prosthetic ones; diseased organs must be removed or replaced by either transplanted ones[23] or by non-organic manufactured substitutes (such as heart pacers and/or stents); blocked arteries must be unblocked (e.g. with angioplasty), and so on.

A notable historian of medicine, Roy Porter (1996, 96) has put this point well:

> surgery was human engineering; as with car maintenance, one peered under the bonnet and repaired faulty parts. Nowadays, transplant surgery permits, for the first time, replacement of parts that are beyond repair. Mechanical and reductionist approaches found their culmination in spare-part surgery.

One might even venture to say that modern, high-tech surgery, based on the latest theoretical advances in the various relevant basic sciences, has reached the height of engineering the human body (save in the case of the brain, an organ which resists transplantation) in accordance with the ontological axiom of human-is-machine. It is true that such a body must be fed with drugs to prevent it from rejecting foreign parts; a machine cannibalized from other machines must need greater attention and care than one originally manufactured with all its parts as a functioning whole. The next step in surgery promises even to overcome this limitation as stem cell research could grow new organs from a cell of the individual human body.

Today, surgery has travelled a very long way from its trade origin as barber-surgeons. Far from being disdained, surgeons – in particular, neurosurgeons – have come to stand for the utmost in human cognitive development. A common saying goes: "as clever as a brain surgeon."

Pharmacology

Biomedical pharmacology differs from traditional (that is, before the seventeenth century) pharmacology[24] in at least two main ways: it is reductionist as well as Paracelsian in character. However, it is worth labouring that the reductionist perspective in this domain is embedded in a framework whose parameters are laid down by the ontological *volteface* which considers the naturally-occurring as machine – this can be seen quite clearly in this quotation from John Locke (the famous empiricist English philosopher)'s *Essay on the Human Understanding* (1690):

> Did we know the mechanical affectations of the particles of rhubarb, opium, and a man, as a watchmaker does those of a watch, whereby

it performs its operations, and of a file which by rubbing on them will alter the figure of any of the wheels; we should be able to tell beforehand that rhubarb will purge, hemlock kill and opium make a man sleep.

This section, however, will also consider another aspect of biomedical pharmacology, namely, the implications for psychopharmacology and its understanding of the placebo effect arising from the relationship between matter and mind as articulated by Descartes, which we have already discussed in Chapter 3.

Reductionist in character

Its reductionist approach will be considered in the design and the discovery of drugs based on their chemical structures at the molecular, even nano level, in targeting patients at different levels of biological organization whether organs, tissue, cell, DNA sequences. This section will explore the various stages in the reductionist approach in biomedical pharmacology; however, the stages distinguished are entirely analytical in purpose and not meant necessarily to represent the actual historical sequences, although on the whole they do.

Drugs have long rested on the usual three sources, namely, vegetable, animal and mineral. However, before the biomedical era, these were used as wholes, even though often only parts of a plant (such as its leaves, roots or bark), parts of an animal (such as one of its organs or bones) might be used and were, in general, processed (being dried or roasted, then ground even into powder) before consumption. In particular, plants were greatly relied on. However, this no longer held with the advent of the biomedical era, as a fundamental change in approach occurred in the early nineteenth century when François Magendie in 1809 claimed that the efficacy of a plant remedy rests solely on one particular chemical in the plant which may be called the active ingredient, and that its efficacy is dependent on its availability in a pure form. This began the search for active ingredients in plants known to have curative properties, to isolate these special chemicals from the others which plants also possess in abundance.[25] Magendie, together with Pierre-Joseph Pelletier (1788–1842) soon made good these claims by isolating emetine, the active ingredient from ipecac[26] in 1817. Pelletier and Joseph Caventou (1795–1877) in 1819 isolated strychnine from the bean called *Strychnos ignatii* and quinine from the bark of the cinchona tree in 1820.[27]

The search for and the isolation of active ingredients is nothing but the "no frills, no nonsense, minimalist" approach to therapeutic remedies. The remedy should consist only of that chemical alone which makes an ostensible difference; the other chemical properties of the original plant/animal are *ex hypothesi* discarded as irrelevant or superfluous to requirement. This assumption may then be regarded as an axiom of biomedical pharmacology, which forms the first stage in the reductionist process.

The second stage is reached once the details of the molecular structure of the isolated active ingredients enable chemists to synthesize the molecule *in vitro*. This meant that the plants themselves would be rendered irrelevant or superfluous. Synthetic substitutes are in the long run cheaper to produce *ab initio* from molecules in the lab, than products processed from naturally-occurring ingredients harvested as cultivated crops or from the wild.[28] Moreover, under conditions of mass, conveyor-belt manufacturing in a factory, there would be perfect quality control. Thus, quinine was synthesized in 1944 by Robert Woodward and William Doering.[29]

The third stage consists not merely of producing synthetic organic chemicals, but also modifying their respective structures, with the hope of producing better drugs. This kind of drug-design programme consists of branching, lengthening or shortening the chemical chain, altering the kinds or positions of its components, replacing rings by other cyclic structures. The first success in this biomedicinal chemistry is based on Paul Ehrlich (1854–1915)'s synthesis of arsenical chemo-therapeutics which forms "a transition to planned molecular modification. Inorganic arsenicals had proved toxic to several pathogens, and it was hoped that organic derivatives of arsenic would be more acceptable pharmacologically to the infected host."[30] Ehrlich's neosalvarsan for treating syphilis is one such product.[31]

Today sees yet another stage at work. The screening involved in searching for new therapeutic drug structures is no longer done by hours of patient human effort – it is now a fully automated process. Pharmaceutical companies active in Research and Development (R & D) would have an extensive library of chemical compounds (GlaxoKlineSmith's is reputed to have two million items). A small quantity of each is put into a test tube into which a molecule of a particular diseased cell (called the target) would be introduced. Out of this huge number, only perhaps in a few cases would the target produce a reaction – the scientists would then concentrate on these, which would have to be refined; defects would have to be removed. Fifteen years of modification and testing plus at least one billion US dollars would have to be spent to find one such drug successfully brought to market.[32]

On the further reductionist horizon, two promising radical pharmacological strategies beckon: one is what Chapter 7 has raised *en passant*, namely pharmacogenomics, the other is nanomedicine. Pharmacogenomics, as already observed, involves delivering drugs via targeting the patient's SNPs, that is to say, at the level of those DNA sequences which distinguish the individual genome from those of others. Such therapeutic treatment is bespoke medicine.

Nanomedicine is the application of nanotechnology[33] to medicine. A nano is one billionth of a metre, so small that it is the size of an atom. Nanomedicine is, of course, as yet a gleam in the eye, so to speak – a promised land rather than reality. In 2005, the US government set up the NIH Roadmap's Nanomedicine Initiative, a programme of research which is expected to deliver "goods" within ten years of its establishment. At nanoscale, particles have been discovered to have physical and chemical properties which they do not otherwise possess. Nanopharmaceutics are part of nanomedicine. It is held that drugs at about 100 nanometres or less can deliver medicine to cells in tissues to cure disease, to repair damaged tissues (whether bone, muscle or nerve), to create new structures which can function within cells and tissues, to introduce nano-sized machines or engines to scour blocked arteries, and so on. [34] This new method of therapeutic delivery would be much more efficient than any extant methods as it increases bioavailability (enabling the body to absorb and process the chemical more readily) as it is centred on cell precision.

One last observation about the nature of biomedical pharmacology – apart from its reductionist perspective within the body-is-machine ontological framework – is that biomedical drugs are designed to disrupt fundamental biological functions. One of the clearest examples comes from contraceptives. Female contraceptives have been with us from the 1950s and today are probably taken by 100 million women worldwide. Today, there are two main versions, the one with only the hormone – oestrogen, typically ethinyl oestradiol – and the other is the combination birth control pill, with oestrogen and progestin, of which there are several types. The former thickens the mucus in the cervix, preventing the sperm from getting to the egg as well as thinning the lining of the uterus so that an egg, even when fertilized by a sperm, would have difficulty implanting itself there. Oestrogen also inhibits ovulation but by preventing what is called follicular development.[35] Ovulation is a normal biological function in women of reproductive age – women in this age group as a rule ovulate once a month but are prevented from doing so by contraceptive pills should they take them. No male contraceptives are available commercially (for a variety of social reasons) although at least one has been successfully researched and developed. This too works upon suppressing a fundamental male biological function, namely, the production of some 200 million sperms each day by the testes. The pill contains synthetic hormones which instruct the pituitary gland to suppress sperm production. However, this has the effect of emasculating manhood; so another hormone must be introduced at the same time to give back the suppressed testosterone to restore the male patient to normal.[36]

Paracelsian in character

We need first to say a few brief words about Paracelsus (1493–1541) who was not only qualified in the medicine of his time, but who also considered himself an expert in alchemy. He was a pioneer in alchemical medicine. Yet in spite of such provenance, he has also been acclaimed a founder of modern medicine. On the surface, such acclamation is preposterous, even while admitting that alchemy is regarded as the forerunner of modern scientific chemistry. The honour bestowed on this late medieval physician has of late, as we shall see, been boosted by developments in biomedical pharmacology and their outcome.

The sub-section above has already shown that biomedical drugs are based on the chemical structures of the active ingredients isolated in therapeutic remedies, then synthesized and even modified. We also have seen that a biomedical pharmacopeia relies not merely on plants and animals (that is, organic chemistry) but also on minerals, particularly metals (that is, inorganic chemistry). Paracelsus held that (inorganic) chemistry must be the basic science for physicians; he himself used mercury (for syphilis, gout, leprosy, ulcers) and antimony (for wounds and leprosy). When accused of using poison on his patients, he responded by saying that all medicines use poison, the only difference lies in the dosage. In *Alchemical Medicine*, he said: "The preparations for Antinomy vary with the diseases for which it is administered. That which is used for wounds differs from that which is applied in the case of leprosy. And so of the rest. To take the same preparation of Antinomy both in wounds and in leprosy would be a serious error."[37]

Paracelsus was ahead of his time, given his own avowal to put medicine on a new and radical footing, although there may be some grounds for denying him the honour of being a founder of modern medicine, because of his obsession with alchemy.[38] On the other hand, there would be perfect consensus to confer that status on Paul Ehrlich, who was awarded in 1908 the Nobel Prize in Medicine and Physiology for his work in immunology. Ehrlich was a pioneer on numerous fronts. Earlier we have already mentioned him for his work in biomedical chemistry, in initiating a programme of modifying the molecular structure of chemicals in drug production. He experimented with chemical dyes to see if these could have therapeutic properties. Like Paracelsus, he too worked on poisons, as arsenic is certainly one of the most potent poisons known to us. However, unlike Paracelsus, Ehrlich was much more optimistic and confident that the drugs one could produce in a biomedical lab (unlike those produced in alchemical ones) would have no ill side effects. Such a type of drug is what he called "magic bullets". These compounds would attract only disease-causing micro-organisms and destroy them, leaving untouched other organisms as well as no harmful effects on the rest of the bodies of patients. He wrote:

> If we picture an organism as infected by a certain species of bacterium, it will...be easy to effect a cure if substances have been discovered which have a specific affinity for these bacteria and act...on these alone...while they possess no affinity for the normal constituents of the body...such substances would then be...magic bullets.[39]

He had some success in finding compounds to treat malaria and sleeping sickness, but his ideal of the magic bullet was not realized till he found some arsenicals, as we have seen, such as Salvarsan and Neosalvarsan, which turned out to be effective treatments for syphilis, thus establishing his reputation as a key pioneer in chemotherapy. His real achievement, however, it is said, lies in inspiring other researchers in their search for magic bullets, such as Gertrude Elion and George Hitchings who, eight decades or so later, were awarded the Nobel Prize in Medicine in 1988.[40]

Ehrlich's concept of the magic bullet, in spite of its ostensible success, turns out to have worrying implications, contrary to what Ehrlich maintained; the most obvious is that magic bullets appear to have some very serious side effects.

Just to take one example – thalidomide was marketed in 1957 as a tranquilliser and painkiller, effective in treating insomnia, colds, coughs, headaches. It was also found to be an effective anti-emetic and, as a result, was prescribed for pregnant women to help curb the symptoms of morning sickness. Unfortunately, it turned out that the drug could cross placenta barriers, affecting foetal development, causing some babies to be born with severe limb defects. When the drug was being developed and later released, medical science did not realize that drugs could cross such barriers. Trials on animals during the process of R & D of the drug did not come up with such side effects. However, in the light of this new data, thalidomide was withdrawn, once the link between it and foetal defects was established. The tragedy involving more than 10,000 children in 46 countries throughout the world prompted tightening of procedures in clinical trials, including tests for teratogenic effects in the foetuses of pregnant animals. However, pitfalls might still be waiting for such improved trials, as shown by the initial efforts to establish the causal link between thalidomide and birth defects:

> In approximately 10 strains of rats, 15 strains of mice, 11 breeds of rabbit, 2 breeds of dogs, 3 strains of hamsters, 8 species of primates and in other such varied species as cats, armadillos, guinea pigs, swine and ferrets in which thalidomide has been tested, teratogenic effects have been induced only occasionally.[41]

Only when very high doses of thalidomide administered to certain species of rabbits commonly called New Zealand White and primates did the defects show up. This meant that: "In pregnant animals, differences in the physiological structure, function and biochemistry of the placenta aggravate the usual differences in metabolism, excretion, distribution and absorption that exist between species and make reliable predictions impossible."[42]

Given this and similar experiences with drugs,[43] today Ehrlich's concept of the magic bullet is tarnished and enthusiasm somewhat diminished. Instead, there is increasingly open acknowledgement that practically any substance – even sugar and salt – may induce diseases such as cancer, if taken in sufficiently large doses, as well as a new found enthusiasm for the Paracelsian dictum that all medicines are, indeed, poisons; the only difference between remedy and plain poison lies entirely in the dosage. A version of this dictum is now commonly cited: any drug with beneficial effects is bound to have some serious side effects or no drug with beneficial effects would/could be without.[44] This is even called the first law of pharmacology.[45]

However, one must accept that these side effects would/could generally become manifest only in the long run, long after the drug has been certified safe and released into the world outside the lab to be administered to real patients who take the medication.[46] In that sense, today's biomedical pharmacology implicitly accepts that patients turn out to be the actual guinea pigs, not mice or primates, or even those who volunteer or are paid for taking part in clinical trials. (This implicit acknowledgement in turn implies that Random Controlled Trials have limits as a gold standard, and that therefore the concept of RCT itself may be inherently defective.) Furthermore, medical scientists and doctors today also readily acknowledge that as biomedical drugs are designed to disrupt fundamental biological functions in order to procure therapeutic efficacy, serious side effects are therefore bound up with drug design and construction.[47]

Psychopharmacology

In Chapter 3, we raised Cartesian dualism – that mind and body are two distinct and different substances – as an attempt by Descartes to make room both for the modern scientific project as well as for theology. While science studies the body, theology looks after the soul (or mind as surrogate). This crude division of labour was good enough, in the main, to accommodate the new science. The body is machine; the body is matter, which falls into the domain of physics and later chemistry as well as all the other natural sciences. We have also seen that the matter–machine framework involves a metaphysics and methodology which are reductionist in character.

A medicine constructed within such a framework necessarily centres on the body as matter to account for the ways in which disease or dysfunction could assail such a body, as well as on the remedies to eliminate/ameliorate disease/dysfunction.

Emil Kraepelin (1856–1926) is often credited with being the founder of modern psychiatry. He developed a system of classification for mental illnesses; he distinguished schizophrenia from manic–depressive psychosis, a distinction which is accepted even today.[48] However, his greatest achievement for biomedicine may lie in firmly situating the study of psychiatry within the framework mentioned above – psychiatric disorders are in the main caused by biological/genetic factors, and should be studied using the methods of the natural sciences.[49] He opposed the Freudian approach which considered them to be caused by psychological factors, to painful events in childhood the memory of which has been repressed into the unconscious, or to fanciful traumas not based on reality which have got a grip on the person's unconscious. His entire works posthumously published in 1927 eventually showed the impressiveness of his contributions to the subject. However, although his view was very influential in the early part of the twentieth century, it was eclipsed by the dominance of the Freudian psychoanalytic school. His reputation of late has risen again in spite of the fact that he had been ignored for the greater part of the last century.

He is today credited with being the founder of psychopharmacology although he himself did not use the term but a different one – he called it "pharmacopsychology", which studied the effects of psychoactive drugs on psychological functioning, and the construction of the psyche through the use of such drugs.[50] Psychoactive agents have been around since the dawn of human history; the most common being alcohol, tea, coffee and hallucinogenic herbs.[51] Kraepelin experimented not only with these but also with new ones such as amyl nitrite, chloroform, ethyl ether, morphine on psychological functioning, and so on. However, today's psychopharmacology is differently oriented; the change itself in terminology marks the shift in approach over the last 100 years.

Psychopharmacology involves chemical substances which affect moods, perceptions, consciousness as well as behaviour, working on the central nervous system by altering brain function.[52] Psychopharmacology is, therefore, intimately tied up with the study of the brain – neuroscience has made some great strides forward in the last 50 years or so. We know more about the history of the evolution of the human brain, the various parts which make up the brain and the general functions associated with each of them and some of the mechanisms involved in the execution of these functions.[53] The advance in knowledge is, in no small way, due to the availability of high-tech instruments which enable scientists directly to observe on screen brain activities involved when a person is engaged in a particular activity – such as when shown a certain image, startled by a noise or thinking a certain thought. These technologies include EEG (such a machine registers the electric signals sent out by the brain cells and nerves, in the form of graphs called electroencephalographs – helpful, for instance, in diagnosing epilepsy), MEG (magnetoencephalography, a technology which emerged in the 1970s but has evolved today to become a very powerful tool, incorporating a quantum device to measure small magnetic signals, which reflect changes in the electric signals sent out by the brain cells and nerves[54]) and fMRI (functional magnetic resonance imaging, a recent procedure which uses MRI to measure the small metabolic changes which take place in an active brain[55]).[56]

Since the 1960s, research has established that brain cells (or neurons) function by releasing what are called neurotransmitters, up to 100 different kinds. A neuron sends out as well as receives signals from neighbouring neurons. It is made out of three parts: the cell body containing the nucleus of the cell, maintaining the survival of the neuron via various biochemical transformations indispensable for synthesizing enzymes and other molecules; the dentrites, a hair-like structure enveloping the neuron, acting like electric cables to convey incoming signals to the cell; the axon, which is the nerve fibre responsible for emitting signals to neighbouring neurons. The point of contact between two neurons is the synapse. Put very simplistically, when a signal reaches the axon of a neuron, the neuron emits a chemical substance – a neurotransmitter – which diffuses across the synapse to be picked up by what are called receptors on the ends of the dendrites of its neighbour. Some neurotransmitters are exciters (acting to trigger the receiving neuron) while others are inhibitors which serve to damp down signals in the neighbouring neuron.[57]

Psychopharmacologic drugs act on the brain by attaching themselves to these receptors, either damping down or enhancing the action of the neurotransmitter that binds at the site. Parkinson's disease can be treated with drugs that bind to and block the dopamine-2 receptor. SSRI (selective serotonin re-uptake inhibitor) drugs, of which the commercial brand Prozac is the most well-known, are designed to act on the noradrenaline (norepienphrine) system. Neurotransmitters are, therefore, key players in psychopharmacological drugs. Tranquillizers modify the natural chemicals in the synaptic gap; LSD alters the balance of various neurotransmitters (and in the process can cause sounds to be perceived as colours).

In general, neurotransmitters bind to six or seven receptors. Psychopharmacologic drugs are Cocktail Compounds rather than Magic Bullets (designed to select and hit only one target); as such one particular drug can affect multiple neurotransmitter systems or multiple systems that use the same neurotransmitter. In consequence, such drugs entail serious (undesirable) side effects, although these can be controlled by carefully adjusting their dosage. In a lot of cases, these side effects mimic the symptoms of another disorder – for instance, a drug which is effective in treating depression might cause anxiety. This shows that the brain is being affected in ways which are not part of the deliberate design of the drug and its intended receptor.

The role played by neurotransmitters in psychopharmacologic drugs may be seen as filling the lacuna left by the Cartesian heritage of body and mind as two substances. If these indeed are two very different substances, there can be no interaction between them; yet this seems to fly in the face of common sense. Although Descartes promised to talk about the soul and the relationship between the two, he spent most of his time discussing the body-is-matter-is-machine. Descartes seemed to think that interaction took place in the pineal gland, a small gland inside the brain,[58] which he regarded as the seat of the rational soul, the site in which all thinking took place.[59] Descartes was wildly wrong about the pineal gland and its role as the seat of the rational soul; modern thinking places rational thought in the brain, though not in the small gland lodged within it. However, Descartes departed from the older view that the heart was the seat of rational thought. One could perhaps cheekily claim that the remit of today's neuroscience to reveal rationality in terms of brain cells and their functioning could be said to be a vindication of sorts of the Cartesian approach.

However, apart from the contribution of neuroscience, there is another foundational aspect of psychopharmacology which warrants a brief discussion, based on the principle of classical (Pavlovian) conditioning[60] or operant (Skinnerian) conditioning,[61] which is a variant of the former. Their basic idea is this and constitutes what is called behaviourism: to understand another's behaviour and in turn to affect/alter it, one does not have to resort to the person's inner thoughts. Pavlov discovered his behaviourist techniques from experimenting with dogs. Skinner refined them by relying on rewards and punishments to bring about a desired alteration in behaviour. He used his methodology to analyse behavioural changes in animals. As a result of research initiated under the aegis of both forms of conditioning, many animal models were developed to screen potentially useful drugs. Research found, for instance, that the anti-psychotic drugs interfered specifically with certain conditioned emotional responses, anti-anxiety drugs removed the response to punishment, while stimulant and depressant drugs altered general levels of activity. Such understanding helped psychopharmacology to manufacture drugs targeting the connecting links between brain pathways and human behaviour. One dramatic example is the drug chlorpromazine[62] (a variant of an antihistamine compound called promethazine, discovered in the early 1950s), used to treat schizophrenia. Although this drug is not a cure, nevertheless, it permits many patients to lead a relatively normal life instead of being confined for the remainder of their lives to institutions either in strait-jackets or under severe sedation. This drug is regarded as a Pavlovian de-conditioner. The researcher first exposes the individuals to 30 presentations of a bell; then the bell is paired with an electric shock. Individuals who are normal develop no conditioned response. This phenomenon is called latent inhibition. In contrast, patients suffering from schizophrenia lack latent inhibition, that is to say, they do not learn that the bell is "safe". Such patients, when administered with chlorpromazine, would exhibit latent inhibition.[63]

In the light of the above, the presuppositions of psychopharmacology may be spelt out as follows:

1. Mental illnesses and disturbances of one kind or another ultimately progress to undesirable behaviour, considered to be problematic to the patients, their relatives/friends or care professionals and/or members of the public in general.
2. Such undesirable behaviour can only be controlled by psychopharmacologic drugs, or by therapies such as electro-convulsive therapy (ECT), not by psychotherapies of one kind or another (such as psychoanalysis, counselling, and so on).
3. Such drugs are effective in altering behaviour because they influence the brain and its functioning, helped by operant conditioning. By altering brain pathways, behaviour (including thoughts and emotions) can be affected.
4. In other words, this means that the line of causation runs from matter (the chemicals) in the drug to matter (the chemical/electric/magnetic forces) in the brain. Non-material interventions (such as talking, sympathetic listening, being shown another way of looking at a situation or how to handle a situation) can have no or little material effects. The preferred route is the impact of (inert) matter upon (inert) matter; such impact, in the context of the brain and its neural pathways, brings about changes at the level of behaviour, moods and states of consciousness. This illustrates an aspect of causation sometimes called the billiard ball model, an aspect of Humean causation, which will be explored in greater detail later in Chapter 10.
5. Psychopharmacology was developed not only within the framework of neuroscience but also of an approach in psychology based on Skinnerian operant conditioning, if not Pavlovian conditioning. As the latter itself was developed using animal models, whatever changes in behaviour and emotions obtained would have nothing to do with language and meaning, as animals do not possess language in the way humans do. In other words, the effects of drugs and of conditioning would occur entirely at the physical level. Mind (states of consciousness) could be said to be epiphenomenal rather than interactionist, as Descartes might have implied. The main difference between interactionism and epiphenomenalism as solutions to Cartesian dualism is this: under the former, mental states can bring about changes in physical states just as changes in physical states can

produce changes in mental states;[64] under the latter, the causal direction is only from physical states to mental states, but not *vice versa*.[65] This would account for why biomedicine has chosen to go down the route of psychopharmacology and is dismissive of other forms of psychiatric interventions which appear to adhere to interactionism rather than epiphenomenalism regarding Cartesian dualism.[66]

The Placebo effect

We need now to turn our attention to a matter which is not in keeping with epiphenomenalism, the philosophical framework within which psycho- pharmacology is embedded. This involves a phenomenon commonly called the placebo effect.[67] For the moment, let us just rely on a simple, seemingly innocuous definition of "placebo" in the medical context – placebo (the word itself literally means "I shall please") refers to any medication which is inert but could have the effect of ameliorating the patient's symptoms. The term was and is generally intended to have abusive connotations;[68] it means that the treatment is a sham and a scam. The thing was (but still is in some quarters) considered to be nothing but a piece of charlatanism, as desperate patients expected their doctors to prescribe them something to relieve their condition while their equally desperate doctors could do nothing apart from prescribing what they knew would not be of much use, except as consolation.[69] For instance, in the era before modern pharmacological drugs, doctors might have bled a patient – not because they believed that it would be relevant to diagnosis of the patient, but because bleeding was an expected form of treatment.

The subject for the first time attracted serious scientific consideration in the 1950s. Towards the end of World War II, an American anaesthetist, Henry Beecher, ran out of morphine on the battlefield in Europe. In desperation, a nurse injected the wounded soldier who was being prepared for an operation with a saline solution instead. To Beecher's immense surprise, the patient behaved exactly as if he had been injected with morphine. Beecher repeated the trick as morphine continued to remain unobtainable; each time, he obtained similar results.[70] After the war, at Harvard University, Beecher was able to study the placebo phenomenon seriously, so convinced and impressed was he by the power of placebos. His work and that of other medical colleagues began to convince part of the scientific community[71] that this is an area worthy of systematic research. However, before the phenomenon could be taken really seriously from the standpoint of scientific methodology, the studies purporting to demonstrate it must rule out at least the following: (a) self-limitation – as many illnesses run a short course, patients would get better anyway without medical intervention of any kind, (b) regression to the mean – although chronic illnesses do not "go away" by themselves without medication, nevertheless, they do wax and wane. They get worse for a bit and then get better for a bit. If medical intervention occurs during a period when the condition is at its most severe, to be followed by improvement, then one could be led to infer that the intervention has brought about the improvement. But this would be fallacious reasoning, a case of *post hoc, ergo propter hoc*.[72]

However, the pioneering studies were also methodologically improper as they were carried out without including a control group which received no treatment of any kind.[73] The improvement amongst a third of the patients who received the placebo – as reported by Beecher[74] and others was, in the absence of a control group, not meaningful as the improvement could well have occurred even without a placebo.

Later studies have remedied these methodological flaws. Hence today, there is a reasonably large established corpus of research which cannot be faulted along such methodological lines, establishing that the placebo effect is real – that is to say, that something pharmacologically inert can produce real effects of improvement on the sick body;[75] it is, therefore, not due to the initial malingering of patients, to their lying later about improvements in their conditions, to the practitioners being plain mountebanks. On the contrary, the effects observed cover certain matters which were, until of late, considered to be either impossible or counter-intuitive. The placebo effect holds as follows:[76]

1. It is not merely based on the reports of the subjective experience of the patients (such as "I feel better/less depressed/less pain" and so on).
2. It is open to objective observation and certification by doctors/ researchers in double-blind trials.
3. It can be measured, providing quantitative data for comparison before and after the medical intervention.
4. It covers not simply medication in the form of drugs/pills, but also medical interventions in the form of surgical procedures.[77] Nor does it involve only cases of anxiety/depression/pain, but also ulcers[78] and even cancer.[79]
5. It occurs not only with interventions which are considered to be unorthodox or even forms of quackery (such as acupuncture, herbal remedies) but also in the case of pharmacological drugs and – as

already mentioned above – even of surgical operations. In other words, the effect is observed to occur with medication which are not inert but has known pharmacological properties.[80] (In view of this, we shall, later, have to redefine the original tentative definition given earlier.)

6. To date, not all (though a very large range of) medical conditions have been shown to correlate with the placebo response; in cases which do, researchers have known since the 1970s that patients in general are susceptible to the effect. In view of the near universal capability on the part of patients to display the placebo response at one time or another, the effect cannot be correlated or explained by simple reference to the personality, and so on, of individual patients.

Studies have found that the placebo effect obtains differentially with regard to the following features:[81]

- Colours of pills – yellow ones are more efficacious as anti-depressants; red pills perform better as stimulants; green pills are for reducing anxiety; white ones (antacids) are ulcer-soothing.
- Forms of delivery – capsules are more efficacious than pills; injections are even better than capsules.
- Quantity – (in the case of pure placebos) medication taken say three or four times daily are better than those administered only twice.
- Brand of medication – generic medicines are less effective than those branded with a well-known trademark.
- Names – trade names perceived to be sexy or powerful produce greater effect than dull and mundane ones. ("Viagra" is an instance of such a brilliant marketing feat.)

In the light of more recent research since the 1950s into the placebo effect mentioned above, we now need to return to the question of its definition. We have originally said that a placebo is any medication which is inert but could have the effect of ameliorating the patient's symptoms. This turns out to be too narrow. A more adequate definition may run as follows: under certain circumstances (which include the belief on the part of the patients that they are receiving a powerful medicine which could do good as well as the doctors themselves letting the patients know, either implicitly or directly, that they are enthusiastic and positive about the treatment), a placebo is (a) a medication which may be pharmacologically inert, or indeed (b) a medication which is not inert, but pharmacologically active, or (c) any medical or surgical intervention, which unbeknown to the patient, is a "pretend" intervention. These, nevertheless, could have the effect of ameliorating the patient's symptoms in a way which can be objectively observed and measured.

However, even this more comprehensive definition does not cover an extremely significant result which has become available since 2003, when researchers, Benedetti et al., mounted a trial which involved no placebo control group, as no placebo was used, but only morphine. It was just that one group was told by the doctor(s) that they were getting the painkiller, openly administered; the other group was not told that they would be getting the analgesic, although as a matter of fact, the morphine (same dosage, at the same interval as the first group) was delivered by a hidden computer-controlled pump with no doctor or other medical staff in the room. In other words, while patients in the first group were made aware of the treatment, those in the second were not. The second group reported less pain relief than those in the open injection group; this suggests that awareness – and therefore, expectation based on awareness – appears to make a difference to the outcome. Yet, no inert substance or sham procedure was involved in this study. So, should the difference in outcome still be called the placebo effect? Should that term be confined to the narrower context, where inert substances or sham procedures are used, and should the kind of outcome which occurs in the Benedetti study be called "meaning response"?[82]

Generally, this author considers terminological issues not to be crucial, provided everyone using a term knows precisely how it is to be used. After all, as long as no substantive issues are conjured away, the label *per se* is not germane to the understanding of the phenomenon under study. Koshi and Short (2007, 15) imply that for ethical clarity, we should perhaps apply a different term to characterize the situation mentioned above. However, they do say (2007, 16) they would leave it to future research and discussion to settle the matter.

Let us stick to the term "placebo effect" to cover the whole range of phenomena revealed so far as a result of systematic study. Regardless of whether a new term is coined or not, there is one substantive issue which requires some attention, namely, what kind of explanation may be given for the phenomena which have emerged. Up to now, little (compared with other areas of investigation) has been systematically investigated and hence little is known about how the so-called placebo effect occurs, except for the consensus that it is very likely to be part of the body's own healing processes and mechanisms. As for more precise speculations,[83] these may be distinguished in terms of three distinctive philosophical frameworks in which psychosomatic medicine may be understood: materialism/behaviourism (body), semiotics/meaning (mind), transcendence of the Cartesian mind/body dualism.

Cartesian dualism has left a difficult heritage both for philosophy as well as for psychosomatic medicine. It officially endorses interactionism – namely, that body can affect mind as well as mind body. However, as we have seen, Descartes himself was driven to postulating that the interaction occurred in the pineal gland, which everyone agreed could not even begin to render a satisfactory account of the mechanisms involved. So, some philosophers and scientists opt(ed) to privilege body/matter over mind, to concentrate on the body's physical mechanisms. We have already seen that, under epiphenomenalism, the causal arrow only works from matter to mind, but not from mind to matter. But epiphenomenalism obviously cannot begin to do justice to the placebo effect, as the latter appears to be a case of the mind having an effect on the body (matter). One obvious option is to study the placebo effect exclusively from a physical perspective, to focus on the body's mechanisms –the immune system, brain function, neural receptors, on chemical substances such as interleukins (e.g. cytokines, IL-1), prostaglandins, as well as on principles of conditioning, and so on. [84] This, as we have seen, is a reductionist approach[85] – as scientific advances are made, ultimately all psychological explanations may be translated without residue into physical/chemical terms. In the meantime, we may still invoke psychological talk – but as mere stop-gap. This would be in keeping with the biomedical presupposition that the body-is-machine. The kind of materialism it adopts may no longer be crude materialism but a philosophically more sophisticated type – namely, eliminative materialism.[86] However, philosophically speaking, materialism/behaviourism – whether of the cruder or more sophisticated varieties – is not without severe difficulties.[87]

Another strategy is to privilege mind over matter, that is, to build belief/meaning on the part of patients and doctors into the very essence of the medical/therapeutic context (Moerman 2002, 20):

> It is...apparent...that what people know and understand about medicine, what they experience about healing, what healing processes *mean* can also enhance both autonomous and behavioural healing processes. Meaning can make your immune system work better, and it can make your aspirin work better, too.

Human consciousness is mediated via language; language articulates beliefs. Some beliefs may have powerful meanings and symbolic significance for certain of their adherents. For instance, Moerman (2002, 78) cites a very large study of 28,169 Chinese-American adults, matching them with nearly half a million randomly selected controls of white people – all living in California. If the Chinese-Americans had a combination of disease and a birth year which Chinese astrology considers to be ill-fated, these died significantly earlier (1.3–4.9 years) than the "white" controls with the same disease (which ranged from lymphatic cancer to lung diseases such as bronchitis, emphysema, asthma). This difference is put down to "the strength of commitment to traditional Chinese culture." Moerman (2002, 134) sees two separate sets of phenomena, one called "the meaning response" and the other the placebo response. These two sets partly intersect – in the intersection is found the kind of placebo effect one is engaged with in the investigation of psychosomatic medicine. The phenomenon associated with the Californian study – which involves commitment to Chinese astrology – belongs to the general meaning response but does not fall under the placebo effect; similarly, there are some placebo effects which take place, for instance, under conditioning, which fall outside the realm of psychosomatic medicine. Another way of putting it is to say that, according to Moerman (Ibid.), "the placebo effect in the strict sense is only a special case of the meaning response."

A very important word of caution is called for here. This account of Moerman's view should not be understood as a crude form of Idealism which attributes reality only to mental events and considers the physical world to be totally dependent upon mental images and subjective experiences. However, beliefs (bearing meanings), by their very nature can be anchored either in physical reality or not – for instance, one can believe that the earth is round or that it is flat, that God exists or that an equally powerful countervailing force also exists. Either of these two conflicting beliefs in any one cultural setting can have powerful meaning and significance for their respective adherents.

There is no doubt that beliefs, whether objectively warranted or not, whether consciously held or not, do affect one's behaviour and the body's functioning, for example, the brain and/or the immune system. Furthermore, the placebo effect may not even be mediated via explicitly held beliefs about the efficacy of the drug prescribed (such beliefs inspire expectation that the drug would work), as it may even be prompted by sub-conscious/subliminal cues that the patient has picked up, such as even the fact that the doctor is wearing a white coat with a stethoscope hanging round his neck.[88]

Obviously, the explanatory model advanced by Moerman to account for the placebo effect could readily account for the outcome of the Benedetti 2003 study. At the same time, the model would open the gates to other beliefs and the meaning embedded in them (not merely about a well-established powerful pharmacological drug such as morphine). This in turn would make psychological therapies of one kind or another, such as psychoanalysis (the most controversial and criticized), equally respectable. Does it matter if Freud in the end discovered that his patients were not sexually abused in their childhood and that such reports by them were mere sexual fantasies? Freud did not change his own theory as a result of this momentous discovery; he has often been vilified for it. But why should he? Perhaps Freud had anticipated that beliefs and the meanings embedded in them can produce powerful placebo effects.[89]

However, if the Moerman model is adhered to, it would be yet another small step to people who "peddle" positive thinking, who claim that mind over matter via beliefs is a powerful tool which can unleash potential in individuals and help them achieve what they most want to achieve.[90]

The Moerman model has nothing much to say with regard to the issue of how beliefs which belong to the domain of mind could in the end produce effects at the biochemical level, in the realm of the body.[91]

The third approach rejects Cartesian dualism as well as reductionism, opting neither for body over mind or mind over body. Instead it argues that the human being is unique: the human being is at once both mind and body. A corpse is a mere body but a living human being is what we call a person. A person necessarily possesses mind and body, or more accurately, both mental and physical characteristics. We cannot locate a person except through their body; but in locating the body of the individual person, we have also located that individual's mind. The person's mind is not a free-floating substance separate from their body. Mind cannot exist independently of body; a person's body is also where their mind operates.[92] When a person's mind ceases to function, what remains is mere body – a corpse or "vegetable". Hence, according to this philosophical perspective, the concept of a person (which embodies both physical and mental attributes) is primitive – mental attributes cannot be irreducibly explained in terms of physical attributes and *vice versa*.[93] As this view of the person is grounded firmly in both the biological nature of such a being as well as in the culture-language-belief meaning system of the society to which the individual belongs, the placebo effect can readily be understood to occur within such a unique being. According to Brody (1980, 95):

> no being can be *necessarily* both a biological and a cultural entity without the cultural features influencing the biological ones and vice versa (as the interplay between cultural and biological evolution illustrates). By this view, the placebo effect, in which participation in a specific cultural context produces changes in bodily condition, becomes an expected and understandable, rather than anomalous, finding.

The placebo effect is not a simple but an immensely complex and challenging phenomenon to study and understand in all aspects. We have dealt at some length with this issue because it demonstrates neatly the main thesis of this book – namely, that medicine (or indeed any scientific activity) cannot be fully grasped or understood without tracing it back to its philosophical roots. The problems it poses – and the very solutions to such problems – may well be coloured by such origins. Indeed, what phenomena are considered normal/ believable/acceptable and what abnormal/anomalous/unbelievable/unacceptable may be determined by boundaries laid down by the philosophical framework – presupposed by the very activities said to be scientific. The philosophical framework itself, unfortunately, remains invisible, in general, to those who pursue these activities in the name of science. As a result, scientists are mistakenly led to believe that science has nothing to do with philosophy. Indeed the two are commonly considered to be mutually exclusive – to do science, one must leave "obscurantist" philosophy behind. However, this could not be further from the truth. Furthermore, once grasped, the realization may well invite radical rethinking about many aspects of biomedicine and its induced technologies, which rest on the patient-is-body/matter-is-machine axiom. As long as the placebo effect is regarded as anomalous (by the more polite or scholarly) or mere quackery (by those with less or little understanding of the phenomenon), it can be either simply noted or dismissed. However, anomalies in the history of science can have the effect of finally overthrowing a theory or framework. Furthermore, it may no longer be so easy to distinguish between quackery or sham on the one hand and the so-called *bona fide* pharmacological drugs and other medical interventions on the other, if the whole therapeutic context is taken into consideration.[94]

Conclusion

This chapter has attempted to establish the following theses:

1. Biomedical technology is necessarily high tech, as it is generated by basic scientific theories which are deeper than the ones they subsume or replace.
2. Surgery and pharmacology, as biomedical technologies, illustrate in detail the "body-is-machine" at work, as forms of ENGINEERING and engineering.
3. Pharmacology, in particular, demonstrates very clearly the various steps in the reductionist process in drug development and manufacture. The discussion outlined also establishes that it is necessarily Paracelsian in character; that harmful side effects are endemic, and that RCT at the pre-clinical and even clinical stages can give no protection against them.
4. The placebo effect brings out very clearly that biomedicine cannot leave philosophy out of the picture; to do it justice in scientific terms presupposes that we get the philosophy right – in which the living human being, who is a person, lives a meaningful life. In doing so, it may well entail calling into question both the philosophical, as well as methodological, framework of pharmacology itself – and indeed, even perhaps of biomedicine itself – as this medicine starts from the axiom of body-is-matter-is-machine.

Part III
Causality and Disease

9

Nosology: The Monogenic
Conception of Disease

This chapter examines a particular sub-category of the aetiological definition of disease, namely, the monogenic account. In turn, it will explore
in detail a particular conception of that sub-category, the infectious-
agent model, setting out the background for its ascendancy as well as
looking at some of the reasons which have sustained its century-long
reputation as a progressive research programme in biomedicine. It will,
however, also critically assess how such a programme copes with the
anomalies which confront it.

What is nosology?

Nosology is just the technical term for that branch of medicine which
deals with the classification of diseases; such a classification in turn
helps towards diagnosis. One can broadly identify four[1] different noso-
logical approaches in the history of the subject. Diseases may be:

1. Phenomenologically defined – this in turn can be divided into historical and contemporary versions. The former rests primarily on (a)
   the symptoms which were reported by patients in terms of their own
   subjective experience of the illness (a pain here, discomfort there,
   headache, dizziness, and so on). Sometimes, this was expanded to
   include what family and friends might have observed about the
   patient; (b) signs, as observed in an earlier chapter, were what doctors
   observed about patients – these data were considered to be objective,
   especially when they were procured via simple instruments such as
   the stethoscope, the thermometer, and so on. Signs were therefore the
   real basis for identifying, classifying and understanding diseases –
   the most brilliant practitioner and theoretician of this approach is

said to be Thomas Sydenham (1624–1689) who provided as accurate as possible natural descriptions of various diseases, just like a botanist giving as detailed, objective and accurate as possible an account of the plants he studied.[2]

The contemporary version would consider this kind of diagnostic approach as at best preliminary and eliminative. For instance, the GP, relying on the single symptom of chronic diarrhoea, would refer the patient to specialists who may then diagnose cancer of the colon, or ulcerative colitis, using up to date sophisticated technology and tests.

The doctors may also rely on what are called syndromes, that is, the simultaneous presence in a patient of a fixed combination of clinical data (or signs). For instance, rheumatoid arthritis is identified in terms of morning stiffness, arthritis in three or more joints, arthritis in hand joints; symmetric arthritis; subcutaneous nodules; positive reaction for rheumatoid factor, and radiological changes in the joints. A patient is said to suffer from the disease if at least four of these seven characteristics are satisfied.

2. Anatomically-cum-pathologically defined – we have seen in Chapter 7 that anatomy was the first biomedical science to be established which was later enriched by pathological findings. As a result anatomical lesions served to define diseases such as myocardial infarction (commonly called heart attack) which at the time of diagnosis would not have been revealed in the absence of suitable technology. However, today, scans, electrocardiograms, coronary angiograms, and so on, are available for diagnosis; hence, this criterion may be less significant.

3. Physiologically-cum-metabolically defined – the defining criterion may be a single clinical or para-clinical finding. For instance arterial hypertension is determined by blood pressure to be interpreted within the distinction between normal and abnormal levels.

4. Aetiologically defined – this amounts to the disease being defined in terms of its cause. This group then covers the following types of causes: (a) infectious agents, such as bacteria, virus, prions, fungi, parasites, covering diseases such as tuberculosis, malaria, syphilis, and so on; (b) genetics in diseases such as cystic fibrosis, muscular dystrophy; (c) poisons; (d) environmental agents such as in asbestosis. Sometimes, the definition incorporates the aetiological and

anatomical perspective such as in pneumoconiosis (black lung disease which coalminers are prone to).

Prestigious status of infectious causal agents

As can be seen from the above, the aetiologically defined classification of diseases actually covers at least four different sub-groups of causal agents. Yet what has gripped the imagination of the medical establishment more so than the others for more than a hundred years, and still continues to do so, is only one of the sub-groups, namely, that of infectious diseases. What evidence can be cited for making this claim? This includes at least the following:

1. The rise of bacteriology as a biomedical science was pioneered by commonly acknowledged giants such as Pasteur (1822–1895) and Koch (1843–1910), amongst others.[3] Pasteur's work on microbiology not only covered disease-bearing organisms but also those which damaged industries such as sericulture (silk), viticulture (grape vine) and milk (pasteurization). He put the last nail in the coffin of spontaneous generation. In sum, he is commonly said to be the father of the germ theory of disease which covered not merely harmful bacteria but also viruses as well (as shown by his work on rabies including producing a vaccine against the disease). Koch discovered the cause of anthrax as well as that of tuberculosis – the tubercle bacillus. Also famously associated with him are his four methodological postulates which constitute a so-called gold standard for determining etiologically defined diseases in terms of infectious agents.[4] Within two decades (1881–1899), an impressive number of germs were found for a variety of diseases including cholera,[5] diphtheria, typhoid, tetanus, plague, rabies, tuberculosis. These discoveries were aided and abetted by the use of industrial dyes, developed by German chemists, which made the bacteria visible as well as allowed different bacterial species to be distinguished via the technique of staining.[6]

The efforts of Pasteur, Koch and other pioneering scientists put paid to the miasma theory[7] as well as the humour theory of disease;[8] the latter had held sway for two thousand years, if not more – ever since the time of ancient Greek medicine as propagated by Hippocrates, and later Galen, right up to the nineteenth century. The four humours, as already observed, were fluids. On the other hand, the infectious agent is a discrete and distinct entity; furthermore, anatomy/pathology as well as surgery could locate disease in specific organs and their lesions. As a result, the term "solid" medicine came into existence.

2. The rise of the science of bacteriology roughly coincided with the arrival of new therapeutic treatments which started to come on stream replacing traditional therapies (such as venesection), generally acknowledged to have been useless. Pasteur had pioneered a successful vaccine against rabies. Paul Ehrlich had advanced the idea of the magic bullet; in 1909, he and Sahachiro Hata demonstrated that Salvarsan, their arsenical compound, could kill the spirochete of syphilis without drastic toxic effects such as killing the patient. Admittedly, Koch's vaccine, called tuberculin, was a distinct failure; an effective treatment had to await the arrival in 1946 of streptomycin, an antibiotic, which was part of the mid-twentieth century version of Ehrlich's magic bullet when Fleming's chance discovery of penicillin (in 1928) was translated into mass manufacture through the efforts of Florey and Chain during the Second World War. This story of the rise of antibiotics ushers in an unprecedented era of pharmacological and therapeutic success in biomedicine.[9] As a result of the rich theoretical crop of discoveries of infectious agents as well as brand new effective drugs in treating the diagnosed diseases, it is not a wonder that "the germ theory of disease" has become seared into the consciousness of the medical establishment as well as that of lay people as a "golden age." [10]

3. The Nobel Foundation began to award its prize in medicine (and physiology) in 1901. In 1905, Koch was the recipient for having identified the tubercle bacillus as the cause of tuberculosis in 1882; exactly a century later in 2005, Warren and Marshall were the recipients for having identified *Helicobacter pylori* as the cause of peptic ulcer.[11]

During an interval of nearly one hundred and ten years, the award has been given on 16 different occasions to 27 scientists for discovery of the causal agents of infectious diseases, their mechanisms and/or their treatments: von Behring (1901, serum therapy especially against diphtheria); Ross (1902, malaria); Koch (1905, TB); Lavaran (1907, protozoa); Nicolle (1928, typhus); Domagk (1939, antibacterial effects of prontosil); Fleming, Chain and Florey (1945, discovery and manufacture of penicillin); Theiler (1951, yellow fever); Waksman (1952, streptomycin against tuberculosis); Enders, Weller and Robbins (1954, "for their discovery of the ability of poliomyelitis viruses to grow in cultures of various types of tissue"); Rous (1966, tumour-inducing viruses); Delbrück, Hershey and Luria (1969, for replication mechanism and genetic structure of viruses); Blumberg and Gajdusek (1976, respectively jaundice virus and kuru virus); Prusiner (1997, prions); Marshall and Warren (2005, peptic ulcer); zur Hausen (2008, the human papilloma viruses) as well as Barré-Sinoussi and Montaigner (2008, the human immunodeficiency virus or HIV).

By comparison, surgery over the same period was singled out only thrice: Kocher (1909, physiology, pathology and surgery of the thyroid gland); Moniz (1949, who shared the prize with Hess for his clinical application of neurophysiology[12]); Murray and Thomas (1990, organ and cell transplantation).

What is more germane is perhaps the number of prizes given for work in genetics (and its closely related subject, molecular biology), another sub-group of aetiologically defined diseases. There are none awarded for work based on classical Mendelian genetics; however, since the emergence of the second genetic revolution in the last century, nine occasions may be identified, involving a total of 21 scientists. These are: Crick, Watson and Wilkins (1962, molecular structure of nucleic acid); Holley, Khorana and Nirenberg (1968, interpretation of the genetic code and its function in protein synthesis); Arber, Nathans and Smith (1978, restriction enzymes); McClintock (1983, mobile genetic elements); Tonegawa (1987, genetic principle for generation of antibody diversity); Roberts and Sharp (1993, split genes); Brenner, Horvitz and Sulston (2002, genetic regulation of organ development and programmed cell death); Fire and Mellos (2006, RNA interference – gene silencing by double-stranded RNA); Capecchi, Evans and Smithies (2007, principles for introducing specific gene modifications by use of embryonic stem cells). Strictly speaking, only the 2007 award has direct implications for medicine and medical treatment; the others fall more into theoretical understanding of genes and their expression which may have eventual implications for medicine.

Clearly, the Swedish Nobel selection panel is/was in close touch with the medical establishments in countries which count(ed) on the global medical stage; as a result, one could safely say that its choice in general reflects the international consensus of medical science about the exceptional merits of the works honoured, the esteem of the scientists singled out for their contributions, and crucially from the point of view of this discussion, the high status enjoyed by the particular sub-branch of aetiologically defined diseases, namely, infectious causal agents.

Monogenic conception of disease

There are two general points which should be made in connection with the aetiological standpoint in understanding disease:

1. This was put forward forcefully by Koch in 1901; he said that "diseases are best controlled and understood by means of causes, in particular, by causes that are *natural* (that is, they depend on forces of nature as opposed to the wilful transgression of moral or social norms), *universal* (that is, the same cause is common to every instance of a given disease), and *necessary* (that is, a disease does not occur in the absence of its cause)."[13] It supersedes (a) the very old religious view that the causes of diseases were supernatural in origin, being forms of divine displeasure and whose treatment consisted of prayers or magic; (b) the Hippocratic/Galenic account which although was in terms of natural causes, but whose framework was part of "fluid", not solid medicine. It amounts to a paradigm shift in medical thinking when first introduced by Koch, a shift as profound as the theory of natural evolution and natural selection in the history of biological thought. At the same time, it also focused (at least by Koch and fellow bacteriologists) in the main on external agents[14] as the natural causes of diseases rather than on internal factors such as anatomical or physiological conditions. That is why the aetiological conception of disease is commonly said to put in place a new research programme (a term used by Lakatos 1970 in his philosophy of science). The research programme based on the aetiological standpoint may still involve "progressive" rather than "degenerating" problem-shifts[15] because it continues to yield fruitful results and leads to new discoveries, unlike the latter kind which leads to dead-ends and nothing but growing numbers of serious anomalies. Marshall's and Warren's discovery of *Helicobacter pylori* as the cause of peptic ulcer, a century after Koch's discovery of the tubercle bacillus as the cause of tuberculosis, bears testimony to its healthy status as a progressive programme in medical research and therapeutic treatments.

2. Not only are the causes natural, universal and necessary, it is also the case that the cause of a particular disease is limited to a single factor. Hence, it is also known as the monogenic conception of disease,

of which the infectious-agent theory is a striking and an impressive representative. The source of this could be traced back to Sydenham, but it was Pasteur who secured for it the prominence it has ever since enjoyed. However, although Pasteur made the grand pronouncement of one cause, one disease, it was left to Koch (and others) to complete the task. In 1876, Koch presented his findings on anthrax to members of the Botanical Institute in Breslau; he said that "each disease is caused by one particular microbe – and by one alone. Only an anthrax microbe causes anthrax; only a typhoid microbe can cause typhoid fever."[16] As one commentator (Taylor 1979, 21) puts it: "the final hope and aim of medical science is the establishment of monogenic disease entities."

3. It is also associated with a set of methodological rules or guidelines for ascertaining the cause, sometimes called Koch's postulates or Koch-Henle's postulates. These are:
    (a) The bacteria be present in every case of the disease.
    (b) The bacteria be isolated from the host of the disease and grown in a pure culture.
    (c) The specific disease be reproduced when a pure culture of the bacteria is inoculated into a healthy susceptible host. (This really means animals, those which could be infected with the bacteria.)
    (d) The bacteria be recoverable from the experimentally infected host.

Curiously, although these are meant to be canonical, even Koch himself recognized that not all could be fulfilled. For instance, the first postulate – in the form of "must" – appeared too strong when Koch realized that many carriers of cholera were healthy, that is, stubbornly showing no symptoms of cholera,[17] although his team in India found the bacillus not only in cholera sufferers, but also in the water tanks from which victims had drawn their water. He cultivated the bacillus, then injected it into animals which also remained stubbornly healthy. In reality, this meant that Koch, at best, satisfied only the first postulate, namely, that the presence of cholera (symptoms) occurred in the presence of the bacillus. However, he had not shown that all healthy people (those without the cholera symptoms) do not have the bacillus. This in turn means that the monogenic conception of disease had run into problems even as it was being successfully launched.

Furthermore, the bacterium *Mycobacterium leprae*, identified as early as 1873 by Hansen – which causes leprosy – appears to be the only disease-bearing bacterium not so far to have been cultured *in vitro*. It was only in 1960 that an animal model for infection with this particular bacterium was found to work with the mouse in a limited way *via* footpad inoculation; it was nearly a hundred years after its identification that in 1971 the nine-banded armadillo model became available for full research into different aspects of the disease.[18] Hence, the wording here is deliberate: "be present", "be isolated", and so on, leaving it ambiguous whether they mean "must" or "may". Virology is sometimes even said to have been retarded because viruses do not appear to conform readily, if at all, to these postulates.[19]

However, such drawbacks notwithstanding,[20] these postulates continue to exercise a compelling hold over research in this field, in spite of criticisms of the canons, as we shall see.

With regard to the tubercle bacillus (*Mycobacterium tuberculosis*), which is the cause of tuberculosis, Koch managed to culture it then inject this cultured bacterium into some healthy guinea pigs which indeed did develop the disease from which they later died. Upon their death, he dissected them and found that their lungs contained the expected signs of the disease. He repeated the series of experiments, cultivating the bacteria obtained from tubercular humans as well as other animals, and injecting guinea pigs with this culture which also developed the disease from which they died. Upon dissecting these corpses, Koch found that their lungs, too, showed masses of the tubercle bacillus. At the completion of these sets of experiments, Koch felt he could make his findings known to the medical public on 24 March 1882. The tubercle bacillus satisfied, to all appearances, all four of his postulates. Perhaps, it was this which set the firm basis for regarding them as constituting a gold standard in bacteriology.

It is said that Warren (the co-receiver with Marshall of the 2005 Nobel Prize in Medicine) discovered the existence of the bacterium, later correctly identified as *Helicobacter pyloris*, in the course of his normal work. He noticed an association between the bacteria and gastritis in 135 gastric biopsy specimens he had collected between 1979 and 1982. Others before him had lighted upon its existence in human stomachs, but unfortunately, although thirty attempts were made in 1981 to cultivate it, they failed to do so. However, through a chance delay, Warren's specimens in April 1982 were left to incubate for five days, not the usual two; this time, colonies of bacteria had formed and were clearly visible.[21] This success means that Koch's postulate 2 had been satisfied.[22]

Next, Marshall tried to satisfy postulate 3. He injected four piglets with the cultured bacteria but he found them to be normal sometime after the injection.[23] Instead, he decided to use himself as the experimental guinea pig. He first ascertained that he was free from gastric disease by undergoing endoscopy before swallowing 50 ml of the cultured brew. Nothing happened at first, but between the fifth to eighth days, he felt nauseous, vomited and developed bad breath (a symptom that the bacillus was doing its work in the stomach). On the tenth day, he had an endoscopy done and *H. pylori* was found in the sample. However, he took no medication for it. Although he did develop gastritis, the gastritis did not go on to develop a peptic ulcer, in spite of the infection.[24] Later biopsies revealed that his stomach no longer contained the bacteria. He admitted that it was puzzling and mysterious how his body had got rid of the infectious agent.[25] His experiment was reported in the popular press; the medical establishment waited for the outcome of a more systematic investigation which Marshall and Warren undertook in October 1982, involving the biopsies of a hundred patients waiting for endoscopy as well as the replication of the investigation in Freemantle Hospital. These studies convinced many, but not everyone, of a causal link between *H. pylori* and peptic ulcer. Final and universal conviction only followed the outcome of another criterion of cause, as we shall see later, which has nothing to do with Koch's postulates.

Many have since argued that Koch's postulates are neither here nor there and should best be forgotten, that they should be replaced by other more suitable criteria.[26] Yet the fascination with them remains as they are ideal methodological guidelines, not to be readily abandoned in favour of alternatives, unless absolutely required. This, as we shall see, is not out of irrational attachment to some outmoded criteria but because postulate 1, in particular, is tied up intrinsically with the monogenic conception of disease and its causation. We shall now explore the implications of postulate 1 for the monogenic conception of disease and we shall see that they are inextricably linked.

Postulate 1 and the monogenic conception of disease

Recall that the monogenic conception has two key points:

1. One cause, one disease in general,[27] and one microbe, one disease in particular.
2. The said cause is natural, universal and necessary.

Thesis 1 implies that the model of causation is mono- not multifactorial. As for Thesis 2, one can for the purpose of this discussion ignore its emphasis on the first desideratum (that the cause be natural) and concentrate on the two remaining characteristics, namely, that the cause is universal and necessary.

To say that the cause must be both universal and necessary in character amounts to saying that the cause should meet necessary and sufficient conditions. [28] This, we submit, would at least be a reasonable interpretation to put on Thesis 2 above.[29] In the quotation from Koch's writing cited earlier, it is clear what he means by "necessary", namely, "a disease does not occur in the absence of its cause." The term is no more and no less than what is normally meant – a necessary condition is what is sometimes called a condition *sine qua non*; that is, in its absence, the effect would not occur. For instance, oxygen is a necessary condition for conflagration – in the absence of oxygen in the room/atmosphere, the fire (indeed, any fire) would/could not take place.[30] Regarding "universal", Koch wrote that "the same cause is common to every instance of a given disease." In other words, whenever the disease occurs, the factor said to be the cause will be found; the presence of the cause guarantees the presence of the effect. This is no more and no less than what one means by "sufficient" condition – the presence of tubercle bacillus always leads to tuberculosis. Or to use a slightly different wording: tuberculosis will only occur if and only if the tubercle bacillus is present (in the patient).

Let us test this interpretation in connection with Marshall's and Warren's discovery that *H. pylori* is the cause of peptic ulcers. Is *H. pylori* a sufficient condition of peptic ulcer? To say that it is entails that every person (ascertained by the usual means, such as through endoscopy) who has *H. pylori* in the stomach would have the disease called peptic ulcer. For the moment, let us understand "the disease called peptic ulcer" to stand as short-hand for a bundle of symptoms and signs which patients would experience and report to their doctors and which the doctors could verify for themselves: belching, heartburn, general discomfort in the abdomen, bloating or fullness after eating, feeling pain, sickness, vomiting, sometimes bleeding via vomiting blood or passing blood in bowel movements.[31]

If peptic ulcer disease is understood in the above fashion, then an anomaly arises given that as high as 90 per cent of people known to have *H. pylori* infection do not develop gastritis or peptic ulcers; only ten per cent do in the presence of the bacteria.[32] Marshall himself observed in 1984 that 43 per cent of healthy blood donors in Freemantle, Australia had the bacteria.[33] One can say that roughly 80 per cent of the world's population harbour *H. pylori*, of which only ten per cent would develop peptic ulcer disease (PUD) – large numbers are therefore asymptomatic.

The above evidence appears then not to be compatible with the claim that *H. pylori* is a sufficient condition for PUD as defined in terms of the symptoms/signs set out earlier. To say that the causal factor X is a sufficient condition for Y (the effect) is to say that the occurrence of X guarantees the occurrence of Y. However, in this instance, the presence of X (*H. pylori*) fails invariably to guarantee Y (PUD).

Well, if *H. pylori* is not sufficient for the occurrence of PUD, is it a necessary condition? It appears not. Consider the following evidence: one needs to distinguish between two kinds of peptic ulcer (PUD) sufferers: (a) 90 per cent of peptic ulcers in the world are caused by *H. pylori* infection – peptic ulcers in developing countries are, in the main, caused by such infection; (b) however, this leaves 10 per cent of peptic ulcer sufferers whose illness is not caused by *H. pylori* infection – this group exists primarily in the developed world. Their ulcers are caused by the use of NSAIDs (non-steroidal anti-inflammatory drugs such as aspirin, ibuprofen, naproxen sodium). About 30 per cent of regular NSAID users have one or more ulcers.[34]

In other words, the effect (PUD) can occur even in the absence of the putative cause (*H. pylori*). If *H. pylori* is a genuine necessary condition, then in its absence PUD cannot occur. QED: as PUD occurs in the absence of *H. pylori*, *H. pylori* cannot be said to be a necessary condition for the occurrence of PUD.

We have now tested the interpretation of "universal" in terms of sufficient or necessary condition. Next, let us try another interpretation in terms of what is sometimes called the constant conjunction thesis. This is to say that "universal" means: whenever X occurs, Y also follows. This claim is derived from the Scottish enlightenment philosopher, David Hume; we shall be exploring it in greater detail later in Chapter 10 but, for the moment, we shall here simply and baldly present it as above. However, this claim falls short of being a causal one, as when Y always follows X, the association could be only an accidental one. For instance, as observed in an earlier chapter, thunder always goes with lightning, but one would be wrong to infer that lightning causes thunder; similarly, night always follows day, but again one would be wrong to infer that day causes night. The cause of the co-occurrence of the two phenomena (X and Y) could be another factor Z. In medical research and clinical trials, to be methodologically sound, the experiment should be designed and constructed in such a way as to eliminate confounding – another factor Z could be the hidden confounding factor. Note, however, that

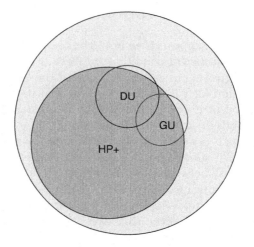

Figure 9.1   Venn diagram of Marshall's and Warren's data

the claim of constant conjunction entails that X is a necessary, but not a sufficient, condition of Y.

It is worth taking a closer look exactly at what Warren and Marshall actually claimed to have discovered in their 1982 investigations. Marshall, in his Nobel Lecture,[35] drew a Venn diagram to illustrate the details, a modified version of which is shown in Figure 1.

His data also showed that 13 out of 13 patients with duodenal ulcer showed the presence of *H. pylori*, that is, 100 per cent; on the other hand, 18 out of 22 patients with gastric ulcer had *H. pylori*, that is, 77 per cent. This shows that Marshall can safely infer that *H. pylori* is a necessary condition for PUD only in the case of duodenal ulcer; in the case of gastric ulcer, *H. pylori* is not even a necessary condition as the bacterium could not be found in roughly a third of the cases. However, in the case of duodenal ulcers, he is also entitled to infer that *H. pylori* is a sufficient condition for PUD. But note that, as far as guidelines for conducting methodologically impeccable clinical trials are concerned, thirteen is surely too small a number as the basis for making such a strong causal claim.

These observations prompt the following query: how can a factor X which fails to be either a necessary or a sufficient condition of Y in all cases be said to be the cause of Y; furthermore, if Y is understood to be no more than constantly conjoined with X, what sense is left of the monogenic conception of disease? Can this conception of disease be "saved"? To see how it can be "saved", let us go through step by step the reasoning involved, using the example below.

Before Captain Cook and other European explorers arrived in Australia, people in Europe held the generalization that all swans were white (P). However, Australian swans were/are found to be black. Given this new discovery, one could make either of two epistemological moves:

1. To admit that P is a false generalization and amend it to read either (a) some swans are white (implying some may not be white), or (b) all swans in the northern hemisphere are white (implying that those in the southern hemisphere may not be white). Or
2. Transform an empirical generalization which turns out to be false into a tautology – this may be called the definitional turn. This is to say that only white swans will be called "swans" – any other bird which is morphologically and in all other biological ways like the white swan would not be called "swans" – call this new proposition P^1. The surface grammar of "All swans are white" (P^1) is the same as that of "All swans are white" (P); however, while P is an empirical though false proposition, P^1 is a true proposition, though it is not an empirical proposition, but an analytically or tautologically true proposition.

In the case of the monogenic conception of disease, both strategies have been used. Regarding the first strategy, one can find careful writers saying not that *H. pylori* is the cause of PUD but, more modestly, that it is the "principal" cause.[36] This cautious claim, however, has the effect of undermining directly the monogenic conception of disease. Hence another ploy may be adopted – to imply that at least one can identify a sub-category of PUD where the monogenic conception can be upheld – this seems to be the way out adopted by Marshall.[37] He implies, at least in the case of duodenal ulcers, that he continues to understand cause in terms of necessary and sufficient conditions. If this interpretation of Marshall is correct, then he can, nevertheless, be said to have revised the claim (the cause of all cases of PUD is *H. pylori*) to the more modest claim that the cause (understood in terms of necessary and sufficient conditions) of all cases of duodenal ulcers is *H. pylori*.[38] Marshall's ploy would be analogous to "all swans in the northern hemisphere are white."

The second strategy was historically prominent and remains tempting in order to protect the monogenic conception from being discredited. This was the route chosen by Koch when he discovered the cause of tuberculosis to be the tubercle bacillus or *Mycobacterium tuberculo-sis*. (This claim is analogous to Warren's and Marshall's discovery that the cause of PUD is *H. pylori*.) In the case of PUD, as we have initially assumed, the disease is defined in terms of a particular set of symp-toms/signs; similarly this was so in the case of what used to be called phthisis or, more popularly, consumption (but renamed as "tuberculo-sis" towards the end of the nineteenth century). "Phthisis" is simply the Greek word for "wasting tissues". Its symptoms included bloody cough, fever, pallor, relentless wasting. When Koch identified the tubercle bacillus and its presence in sufferers with such symptoms/signs, the name of the disease was changed to tuberculosis. When people talk about TB, they appear to have pulmonary tuberculosis in mind; in real-ity, there are many forms, such as scrofula (affecting the lymphatic system, leading to swollen neck glands), TB of the abdomen, the skin, and so on. One may also distinguish between primary and secondary tuberculosis. However, as in all diseases, complexities abound which are hidden by the monogenic conception of disease – in the case of tuber-culosis, infection ranges from no symptoms, to minor illness, to fever, weight loss, to septicaemia, and so on.[39]

The fact that there can be infection without symptoms shows that the bacillus is only a necessary condition of the disease, that is, if "the disease" is identified in terms of a set of symptoms/signs referred to above. If the bacillus is only a necessary condition, it then cannot be the cause of the disease, but only one amongst other causal factors. However, such an acknowledgement would undermine the monogenic conception of disease. To save that conception, the move is made by abandoning the more traditional definition of a disease – in terms of symptoms and signs – and opt for the definitional strategy of defining the disease in terms of a single factor; in the case of infectious diseases, the single factor is the bacterium discovered, be it *M. tuberculosis* or *H. pylori*. It follows from this definitional move that no one could be said to suffer from tuberculosis or *PUD* (let us italicize this word) unless the patient is found to harbour respectively *M. tuberculosis* or *H. pylori* (whether the patient has symptoms/signs or not).

As Bhopal (2002, 103–104) has observed:

> Diseases attributed to single causes are invariably so by definition. For example, tuberculosis is a disease which has many manifesta-tions. It is characterized by a multiplicity of diffuse signs and symp-toms which affect nearly every part of the body, and diagnostic test results which overlap with other diseases. Some diseases, for example,

sarcoidosis, are often indistinguishable form tuberculosis clinically, while the histological finding in Crohn's disease looks very similar to tuberculosis. In some ways tuberculosis is a number of distinct diseases (e.g. pulmonary tuberculosis, cutaneous tuberculosis, tuberculous meningitis), some of which are indistinguishable from other disease. The fact that "tuberculosis" is "caused" by the tubercle bacillus is a matter of definition. In fact the causes of tuberculosis are many, including malnutrition and overcrowding.

At a stroke of the definitional pen, four different things follow:

1. More people in the world could be said to be bearers of the disease (as newly defined), since the number of asymptomatic patients must now be added to the gross figure.
2. It makes sense to eliminate (that is, if therapeutically possible) the cause, as a preventive measure, in the case of asymptomatic bearers, since it is in principle possible that sometime later the bearer may develop the symptoms.
3. Before taking the definitional turn, we can distinguish between the symptoms on the one hand and their cause or causes on the other. For instance in the case of PUD, one can meaningfully say that a cause of PUD could be (a) too much acid secretion in the stomach, or (b) taking NSAIDs, or (c) the presence of *H. pylori*, when PUD is identified and defined in terms of symptoms/signs such as belching, heartburn, general discomfort in the abdomen, bloating or fullness after eating, feeling pain, being sick, vomiting, sometimes bleeding via vomiting blood or passing blood in bowel movements. Under the monogenic conception of disease, *PUD* (written in italics to distinguish it from PUD) is merely identified and defined in terms of the presence of *H. pylori*. As the bacteria in the majority of bearers lead to no symptoms, the bearers themselves would not know that they have *PUD*. Instead, they may be identified in one of two ways: (a) either *H. pylori* in them is accidentally discovered (in the course of investigating something else), or (b) a mass screening programme would have to be undertaken and biopsies made of the stomach wall of the entire population. This, indeed, is a very radical departure from the more traditional practice of patients presenting themselves with what they perceive to be symptoms of illness. It follows from the monogenic conception of disease that probably everyone in the world is a bearer of numerous diseases which, in the individual's lifetime, may never manifest themselves

and of which the individual may never know (unless deliberately screened).

4. The causal conditions for the disease are no longer empirically based but logically secured. The criterion of "the cause" is built into the definition of what constitutes "the disease". Cause and disease are, thereby, analytically linked.[40]

One final observation is called for (before concluding this section) about the complex issues of cause and disease discussed above. This is not a sterile issue, as substantial disagreement exists between those who loyally adhere to the monogenic conception of disease and those who appear to challenge it even today. This kind of challenge is represented by Peter Duesberg in his denial that HIV is the cause of AIDS. One should not be distracted by the fact that Duesberg has, unfortunately, been used by certain governments in their policy of ignoring the therapeutic measures known to control AIDS. Duesberg is a respected molecular and cell biologist at the University of California. A main reason for his claim is that HIV is neither a necessary nor a sufficient condition for AIDS.[41] In this, Evans (1993, 63) concurs with Duesberg:

> HIV is not the single, *necessary* cause of AIDS as infection with another virus, HIV-2, has been shown to result in AIDS in Africa...I also agree that HIV is not a *sufficient* cause as it is dependent on the presence of other latent and/or opportunistic organisms to result in the clinical features. I also agree that HIV does not fulfil the Henle-Koch postulates...[42]

However, Evans disagrees with Duesberg on two counts:

1. In his opinion, the Henle-Koch's postulates are not germane, and as we have already shown earlier even Koch himself had acknowledged that they were not capable of being fulfilled in every case. Evans advocates substitute criteria.[43]
2. He implies that one of these substitute criteria could well be controllability/eliminability, as he believes that an efficacious vaccine against HIV/AIDS could one day be found. When that day comes, then the controversy about the cause(s) of HIV/AIDS would definitively be over.[44]

In the opinion of this book, Evans is partially correct to maintain that Duesberg's denial that HIV is the cause of AIDS should not be linked to the Henle-Koch postulates. Indeed, as earlier argued, postulates 2, 3, and 4 may be superseded by substitute criteria, [45] but not so in the case of postulate 1 (the bacterium is found in every case of the disease), as this postulate constitutes the "essence", so to speak, of the monogenic conception of disease. Rejecting it would be incompatible with the cause of a disease being universal; in turn, if one were to give up this claim, one would be giving up the monogenic conception of disease. Duesberg's overall position appears to imply that he might ultimately be challenging that conception.

Conclusion

This chapter shows that the monogenic conception of disease, in reality, is confronted with evidence which appears to conflict with it. In the face of counter-examples, one strategy of saving that conception has been adopted, namely, to turn an empirical discovery about a causal connection between an infectious agent and a disease into a definitional or tautological truth. The next chapter will explore the causal issues behind the monogenic conception of disease in more detail.

10

Linear Causality and the Monogenic Conception of Disease

In the last chapter, we saw that the monogenic conception of disease, when introduced towards the end of the nineteenth century, was indeed a radical departure. We also raised some anomalies which confronted that conception, as well as some strategies adopted to "save" it. In this chapter, we shall explore the notion of causality which lies behind the conception. This concept of cause has sustained it for more than a century and continues to do so, as a result of which its research programme may still be considered to involve, on the whole, "progressive problemshifts", rather than "degenerating problemshifts", in spite of the problems it faces.[1] That concept of cause is mono-factorial and linear; we shall look at both its strengths and its weaknesses. In the light of such an assessment, it might be plausible to understand the monogenic conception of disease not as enunciating a factual universal truth but as a methodological guideline in medical research into what may count as causes of disease.

Humean roots

From the discussion in the last chapter, one would argue that of the four Henle-Koch postulates, Postulate 1 is really key to understanding the monogenic conception of disease, namely, that every disease has a single cause and that the cause is universal and necessary. In other words, it implies a notion of cause which is mono-factorial and linear. Such a notion is not unique to medicine as it is a conception of cause which could be said to underpin the rise of modern science since the seventeenth century, and which underlies the success first of astronomy, then the (Newtonian) physics, chemistry and other sciences which follow(ed).[2] The most systematic articulation of this conception of cause may be found in the writings of the Scottish Enlightenment philosopher, David Hume (1711–1776).

We have already referred in Chapter 9 to Hume's analysis of cause in terms of constant conjunction or uniformity of sequence; this analysis is conducted within the empiricist philosophical framework. Its strengths and weaknesses in general flow from such a perspective. However, we shall not deal with these in detail here.[3] Instead, we shall focus on issues which are not so often raised in the literature, namely, that it understands the concept of cause to be mono-factorial as well as linear. We shall show how such an understanding stands behind the monogenic conception of disease, in general, and the infectious-agent conception, in particular.

Sometimes, the Humean analysis of cause is also presented as the billiard ball account. This image is very apt, bringing out certain features which are germane to our discussion here:

First: The monogenic infectious-agent conception of disease is "solid" medicine. The cause – whether the infectious agent is a bacterium, virus, or fungus – is something solid, just as a billiard ball is a solid object. Such objects can be seen, measured, "trapped" or caught, although not as easily as a billiard ball. For instance, viruses initially escaped "trapping" by medical scientists. Viruses, too, gave trouble as they could not be seen, unlike bacteria, under an ordinary microscope – researchers had to await the arrival of the electron microscope in the late 1930s, before they could see them. Unlike bacteria, viruses are not cultivable on lifeless media, as they require living tissue – as in a chick embryo – before they can propagate themselves.

Second: When the billiard player pushes his cue against a billiard ball, his push causes it to move, knocking into a second billiard ball, which in turn moves. Motion is imparted from one solid body to another solid body, in an orderly one-to-one fashion. In this context, the original motion comes from the player who stands outside the billiard table; the billiard balls left to themselves would not move. Similarly, the infectious agent, under the monogenic sub-conception of disease is analogously a factor which comes from outside the body – it is an external pathogen which invades the patient's body, rendering the patient sick. The person minus the invasion of his body by an external infectious agent would not fall ill, just as the billiard balls on their own without the impetus imparted to it by the player's cue would not move.

Third: From this image, we can more or less read off Hume's analysis of cause in terms of uniformity of sequence – in the absence of the billiard player and his cue, the first ball would not move, neither would the second. The cue moving the first ball is followed by the ball moving, the moving of that ball against the second ball is followed by the second ball moving. Analogously, the entry of an infectious agent into the body (always) leads to certain "motions" within the body. (Sometimes, these motions may even take the literal form of vomiting, loose bowels or bleeding.)

Fourth: The player imparting motion to the cue, the cue in turn imparting motion to the first ball, the first billiard imparting motion to the second is an illustration at work of (a) one cause, one effect (one germ, one disease) as well as (b) the linear character of cause, which is closely associated with it.

Some anomalies

Yet our earlier discussion has shown that causes of disease rarely, if at all, conform to the mono-factorial conception of cause. Reality differs from this idealized account in at least the following ways:

1. The human body harbours many so-called pathogens. We use the phrase "so-called" because although potentially harmful to the body, pathogens may, nevertheless, never manifest their pathogenic character in the life-time of an individual. In other words, pathogens do not necessarily cause havoc upon invading the body – they could lie inactive for long periods, even within the entire lifetime of an individual. We have seen that 80 per cent of the world's population harbour *H. pylori*, yet 90 per cent of those with such a pathogen do not suffer from PUD, and hence, in the main, such individuals would remain ignorant that, as a matter of fact, they harbour the pathogen.
2. The ability of pathogens to lie inactive means that other conditions must obtain before they become active and produce distressing symptoms in the individual. This means that the existence of the pathogen in the body can at best only be a necessary but not also a sufficient condition. This implies that the dictum one cause, one effect is not satisfied in the majority of cases (that is, if cause is to be understood in terms of necessary and sufficient conditions). Cause is not in this kind of context mono-factorial but multi-factorial.

There are far fewer cases in medical history whose respective causes are genuinely mono-factorial than there are cases whose respective causes are multi-factorial. An example of the former may be Down Syndrome (also called Trisomy 21), whose cause may be traced to the presence of an extra chromosome 21, resulting in the embryo having a total of 47 chromosomes instead of the normal 46 (23 from each parent). It is this extra genetic material which causes the developmental features associated with Down Syndrome.[4] In other cases where a genetic factor is involved, the evidence appears to show that it is only a necessary but not also a sufficient condition for its expression – such is the case, for instance, of phenylketonuria (PKU). This disease is, indeed, an autosomal single gene disease.[5] This abnormal genetic factor results in the body's inability to produce an enzyme necessary to metabolize the amino acid phenylalanine (which forms about 15 per cent of the proteins in most natural foods), to turn it into tyrosine; the amino acid accumulates in the brain, leading to brain damage and the presence of phenylketones in the urine. However, evidence also shows that early diagnosis, together with a diet low in phenylalanine, can stop the occurrence of brain damage.[6]

3. Examples of PKU, as well as *H. pylori* infection cited above, show that we can at best pick out a combination of factors (such as in the former example, single faulty gene,[7] failure of early diagnosis/dietary advice, and hence exposure to a diet high in phenylalanine) which may be said jointly to constitute the cause in terms of necessary and sufficient conditions.[8]

4. Furthermore, the axiom that cause is mono-factorial cannot do justice to a phenomenon in causal matters called synergism or synergetic effect.[9] It refers to the interaction between two things or events whose overall effect is greater than the sum of their respective effects. We may simplistically illustrate it via the proverbial saying: the last straw that breaks the camel's back. For the sake of the argument, let us assume that a pile constituting of 10,000 straws breaks the camel's back. Each straw weighs the same, say, 1g. When 9999 straws have been piled on the camel's back, the animal remains fine and upright. Yet, when the last straw, the 10,000[th], is added to its burden, the camel keels over with its back broken. Why should that single last straw – which weighs no more and no less than each of the other straws – have such a spectacular effect? For the sake of presentational simplicity, let us say here that the 9,999 straws constitute one thing or "event", and the 10,000[th] straw the second thing or event. The first thing on its own did not break the camel's back; neither would/could the second thing on its own produce such an outcome. It took the combination of the two things or events to produce the

dramatic effect, an effect which cannot be explained by simply adding together the respective effects of the two things or events. When synergism takes place, the total effect is greater than the sum of the effects of the parts, so to speak, as the effect is not merely additive.[10] The equation "2 + 2 = 4" is additive; the equation "2 + 2 = more than 4" is not additive and is even unintelligible in terms of strict mathematical rules.[11]

We can return to PKU as an example of synergism at work. The faulty gene on its own may not produce the brain damage; neither does a normal diet (in the absence of the faulty gene) relatively high in phenylalanine. Each of the factors cited in the combination of necessary and sufficient conditions on their own does not bring about brain damage. Yet when the factors happen to combine, then and only then, does brain damage occur. The brain damage is the synergistic effect of all these causally relevant factors.

In the case of drugs, one example which may be cited is the taking of certain medicinal drugs together with alcohol. If an individual were to swallow X number of sleeping pills, this on its own would not kill the person, although the pills may cause some very unpleasant effects. Similarly, if the person were to drink half a bottle of whisky on its own, a very nasty hangover would ensue the next day. However, should the individual opt to swallow the pills at the same time as the alcohol, then death could and would ensue, if the attempt to combine both was not discovered in time by others.

5. Mono-factorial causation and linear causation go hand in hand. Linearity means that of two events which are causally associated, the first is the cause, the second which follows is the effect. As the billiard ball example shows, the second may in turn become the cause of a third event, but the causal arrow always points in one direction only, thus: $X \rightarrow Y \rightarrow Z \rightarrow N$.

6. Mono-factorial and linear causation also ignore what are called standing conditions. Let us return to the camel and straws example. Strictly speaking, it is not merely the synergistic effects of the last straw plus those already loaded on to the back of the camel which account for the camel breaking its back. Other relevant factors include, for instance, the age of the camel, its general health, how hungry and weary it already was, and so on. From the scientific point of view, it is the combination of all these factors – both intervening as well as standing ones – which in the end truly account for

the final effect. One may for certain purposes single out one of these factors as "the cause", but when one does so, one must still bear in mind that the so-called "causal" factor is but one of a long list of factors involved in ultimately giving rise to the effect. Mackie has used the term "inus" conditions to refer to such factors.[12] Each of these conditions, on its own, is neither necessary nor sufficient; however, each is, nevertheless, a necessary component of this particular complex. "Inus" stands for "an insufficient but necessary part of an unnecessary but sufficient complex."

Wulff 1984 cites a clear example to illustrate Mackie's analysis in a medical context. A male patient presents himself with a high temperature and a stiff neck. Upon diagnostic tests, the doctors conclude he suffers from meningitis as his spinal fluid contains pneumococci. His medical records show that his spleen had been removed after an accident; furthermore, he had not been vaccinated against pneumococcal infections following the splenectomy.[13] One could argue, plausibly, that either the splenectomy or the omission of the required vaccination was "the cause" of his meningitis. Each of these two conditions – no more than the invasion of the pathogen – is on its own necessary or sufficient. It is all three factors in combination which cause the meningitis – such a complex of factors constituted a sufficient but not a necessary cause of the illness (as the patient could have developed meningitis in other ways).

"The cause" in different contexts

The discussion above shows that one may single out the omission of the required vaccination as "the cause" should one wish to do so, provided the context is not a scientific explanatory one, but what may be called an attributive or justificatory one.[14] Imagine the patient suing the hospital which failed to vaccinate him against pneumococcal infection after the splenectomy. His counsel would argue that, indeed, it was the negligent omission which was the cause of his meningitis. In turn, the defence for the hospital management could argue that it was only a contributory cause as, if the accident had never happened and the patient had never had his spleen removed, the attack of meningitis might not have occurred.

In other words, an explanatory context is different from other contexts – such as an attributive one, as illustrated above in the case of medical litigation. Science is involved with the former, while ordinary life, law and even medicine under certain specified circumstances are involved with the latter. Of course, medicine is interested in scientific explanations, in establishing laws of nature, which are generalizations about kinds of phenomena; for instance, to discover the generalization that it is the function of the spleen to strengthen the body's immune system so that it could fight off pneumococcal and other infections. However, from an attributive angle, it makes sense to pose the following question: why does this individual (X) succumb to meningitis, and not another (Y) who is similarly matched in terms of sex, age, general health, lifestyle, and so on? Exposure to pneumococcal infection renders X ill but not Y – the difference lies in that X is abnormal compared to Y (and others like Y) in respect of the spleen. X is minus that organ, while Y possesses it. In this kind of attributive context, the doctor is not so much concerned with scientific explanations and generalizations about the body's immune system and immunization, but more about why this particular individual who presents himself has fallen ill and not others like him. The factor which makes the difference is the "abnormal" factor – it is normal for humans to have a spleen, it is abnormal for humans not to have the organ. Similarly, a doctor may pronounce the patient's lapse from his prescribed strict diet to be "the cause" of a bad attack of epigastric pain. The dietary lapse is but one from a set of "inus" conditions which ultimately cause the pain. The justification for doing so is that it is abnormal for such a patient to take food rich in acid, as careful normal avoidance of acid-rich foods means that she would not succumb to epigastric pain. From the standpoint of this particular patient, the failure to adhere to the prescribed strict regimen (the abnormal factor) may legitimately be singled out as "the cause" of her illness.

The distinction between normal and abnormal invoked in an attributive context may, however, in turn lead to a more general explanation. Take this case: why does this patient (X) suffer from PUD and not another individual (Y)? X has been exposed to *H. pylori* but not Y. X compared with Y is abnormal as X has the pathogen, while Y does not. Imagine another scenario: both X and Y have been exposed to *H pylori*, yet only X succumbs to PUD, but not Y. Maybe Y has a genetic component which X lacks and which helps to prevent PUD, just as certain individuals are said to possess some genetic factor or factors in the case of HIV exposure and infection which stand them in good stead against developing AIDS. As such fortunate individuals are less common than those who lack the relevant genetic components, they may be said to be "abnormal" from the statistical point of view.

In other words, the criterion abnormal/normal can be useful in singling out an "inus" condition to be "the cause" under certain specified circumstances and contexts of enquiry.

One must also bear in mind that medical understanding and diagnosis are, ultimately, in the service of treatment; medicine must fulfil its clinical role. In the case of the meningitis patient, his spleen is long gone, vaccination is now too late; as such they may be regarded as "standing conditions" as far as this patient is concerned. Therefore, singling out either of these two "inus" factors as "the cause" appears irrelevant and futile. On the other hand, it would make sense for the doctor to single out the pneumococcal pathogen as "the cause", as it would make a difference to the patient. In today's age of antibiotics, he could be prescribed a relevant antibiotic which would eliminate the pathogen in question. We shall be looking at this criterion of what constitutes "the cause" in greater detail in the next chapter.

Conclusion

We have explored in some detail the notion of causality behind the monogenic conception of disease, tracing it back to its Humean roots, as well as outlining some of the problems it faces. We have distinguished and identified three different contexts which are all relevant to medical pre-occupations: (a) the explanatory/scientific, (b) the attributive, (c) the clinical. From the first perspective, no factor could be singled out as "the cause" as each of the relevant factors which may be identified is neither necessary nor sufficient, each on its own – all the identified factors form a complex set of sufficient ("inus") conditions. On the other hand, from the second and third perspectives, it is legitimate for doctors to single out one of these "inus" conditions as "the cause" by relying, for instance, on the distinction between "abnormal" and "normal". Furthermore, from the third perspective, medicine, as clinical medicine, is pre-occupied with treatment or cure. In this context, doctors may single out one of these "inus" conditions, using the criterion of eliminability or controllability of the disease, as "the cause". Sometimes, it turns out that what is considered to be the abnormal factor is also the factor which one can control/eliminate – for instance, the example of epigastric pain cited earlier shows that the abnormal factor which is the dietary lapse is also a factor which is well within the careful control of the patient.

If the assessment of this and the previous chapter is plausible, this author would like to suggest that it is best not to understand the monogenic conception of disease as enunciating a universal empirical truth – rather, it should be perceived as a methodological guide in medical research. In other words, it amounts to saying: it is fruitful to look for a single factor which could cast important light on the nature of the disease, be it a particular pathogen or a single gene; furthermore, once identified, this factor would encourage research into how to eliminate or control this harmful factor (if a method for doing so does not already exist).

11
Determining "the Cause": Controllability and Random Controlled Trials

In the last two chapters, we have looked at several ways in which biomedicine attempts to determine "the cause" of a disease. These include the monogenic conception of disease in the explanatory context and the distinction between abnormal and normal primarily in the attributive context. This chapter will explore two further attempts in the context of clinical medicine to articulate "the cause" of a disease, namely, the criterion of controllability/eliminability and the notion of the Random Controlled Trial (RCT). It will also argue that these two are closely related as the former's understanding of cause is implicated in the latter; that they are both involved in the notion of experiment; that Mill's methods, in the main, set out the logic of such experimentation.

Controllability/eliminability

Towards the end of the last chapter, we briefly raised this matter. Medicine has a foot in two camps – science and therapy. The former is theory, the latter is practice. As we have seen in Chapters 3 and 8, the goal of the modern scientific project is to discover laws of nature or basic generalizations which in turn would generate new and more powerful technologies, techniques and technological rules with greater efficiency to produce better practical outcomes. In medicine, for instance, scientists study closely the malaria lifecycle; they come to know that the vector is the female of the anopheles mosquito. When it bites an infected human, it then carries the pathogen, which is a parasite, to other individuals by biting them in turn (the Plasmodium, *P. falciparum* being the species posing the most serious threat to human beings).

Zoologists study the biology and lifecycle of the anopheles mosquito itself. Geneticists study the DNA of the parasite as well as that of the vector. As a result, a variety of measures, some more successful than others, have been deployed to either cure the disease called malaria by way of drugs or prevent it from occurring by distributing mosquito-repellent nets for people in infected regions to sleep under to protect them from mosquito bites, by killing the mosquito via chemical sprays (such as DDT), vaccination, and of late even by altering the genetic make-up of the mosquito.

These various methods single out as "the cause" or "primary cause" of malaria either the vector or the parasite itself. As we shall see, these methods rely on the criterion of controllability/eliminability to do the job of selecting one of the numerous "inus" factors to be "the cause".

Ideally, there should be no gap between theoretical scientific understanding of a phenomenon on the one hand and the technology/techniques for controlling/ eliminating the undesired phenomenon (or producing it should it be deemed to be desirable) on the other. However, reality often fails to march hand in hand with what is ideal in two ways: (a) theoretical knowledge fails (at least for the moment) to generate suitable technologies/techniques, (b) an effective technology/technique may exist for controlling the phenomenon, yet science lacks theoretical knowledge/ understanding of why it works in the way it does. In other words, "knowing that" has not led to "knowing how" in the former instance, while "knowing how" has not led to "knowing that" in the second.

A commonplace illustration of the first type of gap today is the common cold. Biomedicine knows infinitely more than it did a hundred years ago about how the various major systems of the body work, what are the nature of viruses in general and even of the common cold virus in particular, and how they work, yet this enormous growth of theoretical knowledge appears not to have produced any cure. At best, there are various "common sense" suggestions for relieving some of its symptoms, such as resting in bed, taking plenty of fluids, and so on, as well as advice to prevent catching one, such as keeping one's distance from those afflicted with it, washing one's hands frequently for fear that they may have been contaminated with the germ, and so on.[1]

A clear example regarding the second type of gap historically is the spectacularly successful methods developed over millennia in the long history of agriculture for the breeding of plants and animals. Yet until the last century, when classical Mendelian genetics and later molecular/DNA genetics were discovered, humankind had no real theoretical understanding about the transmission of hereditary material from parents to offspring; till then, every achievement in this domain, no matter how impressive, was merely the outcome of painful trial and error.

In the history of biomedicine, the relationship between theory and therapy is even more complex than just set out above. In Chapters 9 and 10, we have traced the emergence of the monogenic conception of disease in general, and of the infectious-agent sub-conception in particular. Koch had singled out the tubercle bacillus as "the cause" of tuberculosis; that was in 1882. Fleming accidentally discovered penicillin in 1928 but, as we have seen, the age of antibiotics as the new "magic bullet" in clinical medicine did not arrive till the late 1940s. Streptomycin was used successfully in 1946 to treat tuberculosis. The vaccine called BCG was not used on humans till 1921 and was not widely used until after the Second World War. The traditional methods of "cure", such as rest in sanatoria, remained in place for roughly sixty years. We can see from the brief history above that Koch's championing of the tubercle bacillus as "the cause" of tuberculosis long predated the first effective antibiotic to control the disease. So what justification could one mount on behalf of Koch that he could have controllability in mind when he singled out as "the cause" the tubercle bacillus? Two related pieces of evidence:

(a) Koch, of course, could not have antibiotics in mind; instead, he tried to produce a vaccine against the disease. He did produce one, which he called tuberculin, but it turned out to be a failure. However, this would not necessarily dampen his hope in that direction. Indeed, Koch would have been buoyed by one related spectacular success – Koch himself had demonstrated in 1877 that "the cause" of anthrax was *Bacillium anthracis*; a mere four years later, in 1881, Louis Pasteur as well as two other scientists had independently developed a successful vaccine against anthrax.

(b) Koch would also have been sustained (like all other medical scientists) in his search for a cure/treatment by the general ideological goal of the project of modern science, which is to advance human well-being through controlling and manipulating nature including eliminating disease, pain and suffering – this ideological goal would be a powerful inspirational guide in the pursuit of cures for the infectious diseases identified and distinguished.

In the eyes of doctors, an effective therapy to eliminate/control the disease is of paramount importance, more so than mere knowledge of a theoretical kind about the disease. This is borne out by the eventual acceptance of Marshall's and Warren's claim that *H. pylori* is "the cause" of peptic ulcers. A quick look at that history makes the point clear. In 1984 when they said that they found a strong correlation between *H. pylori* and peptic ulcers, it was not surprising that the medical world ignored the claim, as even they themselves realized that such a correlation did not amount to a causal relationship. They then conducted in 1985 and 1986 a double-blind trial involving a hundred patients with duodenal ulcer and *H. pylori* infection, divided into four groups, each group randomly assigned a treatment as follows: (i) cimetidine, (ii) bismuth, (iii) tinidazole (an antibiotic), (iv) a placebo. The results of this trial after ten weeks were as follows: (a) the ulcer was healed in 92 per cent of patients, with *H. pylori* no longer detected, (b) the ulcer healed in only 61 per cent of patients with persistent *H. pylori*. After 12 months, 84 per cent of patients with persistent *H. pylori* relapsed, whereas only 21 per cent of patients without continuing *H. pylori* relapsed. After seven years, 20 per cent of patients with *H. pylori* infection, but only 3 per cent of patients without the infection, had duodenal ulcers. Some American researchers conducted another trial between 1988 and 1990 with equally encouraging results. These sets of results were impressive but not enough to convince every body, as sceptics claimed that it was bismuth which could have cured the patients of the ulcer and not the eradication of *H. pylori* which had done the trick. This time, Austrian and Dutch researchers conducted further trials which showed that it was the antibacterial drug, in eradicating *H. pylori* – rather than the bismuth and antacids – which were crucial in healing the ulcer.[2] By 1994 in the light of these tests, a consensus emerged which held that *H. pylori* causes peptic ulcers (p), and that peptic ulcers can be cured by antibiotics (q). This consensus was set out in a document issued by the NIH (National Institutes of Health, United States).[3]

In the view of the medical community, what is the relationship between (p) and (q)? Are they separate factual claims which are strongly but, nevertheless, merely contingently related? The NIH Development Panel (1994, 66) stated that:

> the strongest evidence for the pathogenic role of *H. pylori* in peptic ulcer disease is the marked decrease in recurrence rate of ulcers following the *H. pylori* eradication of infection. The prevention of recurrence following eradication is less well documented for gastric ulcer than for duodenal ulcer, but the available data suggest similar efficacy.

Thagard (2000, 62), following the NIH, similarly holds that (q) is simply very strong evidence for (p):

By far, the most impressive evidence that *H. pylori* causes peptic ulcers is the demonstration that eradication of *H. pylori* strongly contributes to the elimination of ulcers and the prevention of their recurrence.

However, they may not be right; the relationship between (p) and (q) is not quite such a simplistic matter, as (q) appears not merely to provide strong evidence in favour of (p). On the contrary, (q), which is about controllability/elimination, constitutes a criterion of what counts as "cause" in (p). As a matter of fact, what Thagard says – see 2000, 61 – appears to be inconsistent with what he says a page later, as quoted above. On the earlier page, he has constructed table 4.1, headed "Criteria for Causation", a schema he has adapted from Evans 1993, 174. This set has ten criteria, of which the 8th and the 9th are germane to our concern here (Thagard 2000, 61):

8. *Elimination or modification* of the putative cause or of the vector carrying it (e.g., via control of polluted water or smoke or removal of the specific agent) should decrease incidence of the disease.

9. *Prevention or modification* of the host's response on exposure to the putative cause (e.g., via immunization, drug to lower cholesterol, specific lymphocyte transfer factor in cancer) should decrease or eliminate the disease.

If X (*H. pylori*) can be eliminated (via in this case an antibiotic such as tinidazole), thereby significantly lowering the incidence of peptic ulcers, then X (*H. pylori*) is "the cause" of Y (peptic ulcers). In other words, controllability/elimination is used to single out one of the factors which form the "inus" conditions for the disease as "the cause" of the disease. This criterion is a potent one – after all, medicine is in the business of relieving pain as well as eradicating disease which causes suffering to humans. While theoretical understanding – what type of bacterium is *H. pylori*, its ability to live in a by and large acid environment in the stomach, its exaggerated release of gastrin in the stomach and its effects on the parietal cells, its increase of acid secretion, and so on – is all very gratifying, but the clinching argument for *H. pylori* being "the cause" of peptic ulcer remains the fact that it can be eradicated and that its eradication in turn leads to the healing of the ulcer and the lesser rate of relapse.

To reinforce the points made above, we shall next explore in some detail the philosophical background of the notion of cause in terms of controllability, leaning on the analysis of the philosopher, R. G. Collingwood (1889–1943). He set out his account in his address to the Aristotelian Society in London, distinguishing between three senses of "cause", of which the second (which he called Sense II) is precisely about controllability. He also specifically mentions that it is this understanding of cause with which medicine is most pre-occupied. Though long, we shall, nevertheless, not hesitate to reproduce here some key passages from that address – Collingwood, 1938, 89–90:

> In Sense II...the word cause expresses an idea relative to human action; but the action is an action intended to control not other humans beings (such as in Sense I), but things in "nature," or "physical" things. In this sense, the "cause" of an event in nature is the handle, so to speak, by which we can manipulate it. If we want to produce or to prevent such a thing, and cannot produce or eliminate it immediately (as we can produce or prevent certain movements of our own bodies), we set about looking for its "cause." The question "what is the cause of an event y?" means in this case "how can we produce or prevent y at will?" This sense of the word may be defined as follows: *A cause is an event or state of things which it is in our power to produce or prevent, and by producing or preventing we can produce or prevent that whose cause it is said to be....*
>
> Suppose someone claimed to have discovered the cause of cancer, but added that his discovery though genuine would not in practice be of any use because the cause he had discovered was not a thing that could be produced or prevented at will. Such a person would be universally ridiculed and despised. No one would admit that he had done what he claimed to do. It would be pointed out that he did not know what the word cause (in the context of medicine, be it understood) meant. For in such a context a proposition of the form x causes y "implies the proposition" x is something that can be produced or prevented at will" as part of the definition of "cause." For in such a context a proposition of the form "x causes y" implies the proposition "x is something that can be produced or prevented at will" as part of the definition of "cause."

The first part of the second long passage requires some special comment, which is to point out that Collingwood was wrong to imply that medicine is nothing but clinical medicine and effective treatments/cures;

that it has no legitimate interest in obtaining theoretical knowledge and understanding about the combination of causal factors leading to diseases. The truth which Collingwood had failed to comprehend and hence ignored is that biomedicine has a foot in both the scientific/ theoretical and the therapeutic domains. Indeed, as our earlier chapters have argued, new therapies in medicine, increasingly, are expected to come from basic theoretical breakthroughs in the biomedical sciences. (For instance, the Human Genome Project and a related basic science such as molecular biology are expected to play this critical role in the near future.) However, for our discussion here of cause in terms of the controllability criterion, we can simply ignore this flawed aspect of Collingwood's one-sided account of medicine.

In any case, he correctly pointed out that this sense of cause belongs to practice, not theory. It belongs to what Aristotle called "practical science", where knowledge is not valued for its own sake, out of intellectual curiosity, but utility. It also sits comfortably with the Baconian view of science in particular (that knowledge is power, that nature is conquered by obeying her) and in general with the ideological goal of the modern project of science (and of medicine), that is, to use science in order to manipulate/control nature to suit our purposes. Therefore, built into this sense of cause, is the idea that cause and effect are related in the way means and ends are related. When we single out a factor as the cause, we are saying in the practical/therapeutic context that it is the means by which we achieve the end of eradicating/ameliorating something deemed undesirable (such as pain, discomfort, death which go with diseases). The relationship between means and ends is about rationality as efficiency, what Kant has called the hypothetical imperative[4] and what Habermas called instrumental rationality.[5]

Collingwood was also correct in pointing out that Sense II must be distinguished from Sense III[6] – the latter belongs to theoretical discourse, what we earlier have called the explanatory context. In this context, we have found Mackie's combination of factors, which he calls "inus" conditions, to be appropriate. Sense II is not interested *per se* in "inus" conditions, or in looking for all the other necessary conditions which may be jointly sufficient for the effect to occur; it simply assumes that they exist. Rather, it is interested *per se* in identifying as cause a factor the manipulation of which would lead to the desired outcome. This explains an earlier observation we have made that the medical community was finally only convinced that *H. pylori* is "the cause" of peptic ulcers when certain antibiotics were convincingly shown to have made a difference to lowering significantly the re-occurrence of the disease through the eradication of the bacterium. We shall see in the next section that this sense of cause is linked with the notion of an experiment, rather than with mere observation.

Random Controlled Trial (RCT)

Today, RCT is acknowledged, universally, to be the methodological gold standard in clinical research for assessing the efficacy of any treatment or innovative intervention in pharmacotherapy, surgery, physiotherapy, diet/nutrition, preventative measures, and so on. However, in its contemporary guise, its history is surprisingly short. After the end of the Second World War, the Medical Research Council in the UK endorsed the protocol designed by Austin Bradford Hill[7] in an attempt to determine the efficacy of streptomycin for treating patients suffering from pulmonary tuberculosis – this RCT and its results became the first to be published, though perhaps not mounted.[8] However, with every socalled new idea, history would reveal earlier prototypes. Without going back too far in time, one would like to draw attention to the efforts of Pierre Charles Alexandre Louis of Paris in 1835 to show that bloodletting was of no value in treating pneumonia,[9] of Ignaz Semmelweiss in Vienna who, in 1847, demonstrated that puerperal fever was both contagious and its incidence was reduced significantly when medical staff conscientiously washed their hands, after visiting the mortuary and touching the cadavers, before tending to patients in the maternity ward,[10] as well as the earlier success of James Lind in 1747 in determining the cause of scurvy amongst (British) sailors and how to treat it. We shall give some details of the latter experiment only. Lind (Silverman 1985, 7) wrote:

> On the 20th of May, 1747, I took twelve patients in the scurvy aboard the Salisbury at sea. Their cases were as similar as I could have them…Two of these were ordered a quart of cider a day. Two others took twenty five gutts of elixir vitriol…Two others took two spoonfuls of vinegar…Two were put under a course of sea water. Two others had each two oranges and one lemon given them each day…the two remaining took the bigness of nutmeg…The consequence was the most sudden and visible good were perceived from the use of the oranges and lemons.

Judged by the methodological standards of today's RCT, Lind's experimental set-up could be said to be highly defective and would be assessed as such by the Cochrane Collaboration,[11] although one must admit that Lind very carefully made sure that he matched the groups of patients according to the seriousness of their affliction, their diet in all other aspects, their accommodation on board. In general, the desiderata of what is sometimes called a "fair trial", [12] minimally include the following:

1. The trial should include large numbers of patients. The reason for this is to eliminate chance. Imagine some patients (matched in the way outlined above) with the same illness. One group was given the new treatment (N), and the other group the standard treatment (S). Five people improved with N while seven improved with S. Can one confidently conclude that N was worse than S? No, because should the experiment be repeated, the results could have been reversed. However, one's confidence would increase should 50 people receiving N improve while 70 receiving S improve, as chance would be less likely to explain this result. Now, should 500 patients receiving N not improve while 700 receiving S improve, then one could confidently pronounce that – for the condition under test – S is better than N, or that N is worse than S. Lind's experiment involved only 12 patients.
2. To reduce bias, RCT demands that patients be randomly assigned to the experimental group and the control group(s). Lind failed to do that. (Today, random assignment could be determined by a computer programme, using a suitable algorithm.)
3. To eliminate the placebo effect, the treatments should be double-blinded (save in situations where this is impossible); neither patients nor medical staff would know which group is receiving which treatment. This information and the results of the experiment would only be known to another body of researchers who have no contact with those engaged in the direct administration of the treatments under trial. Lind also failed to comply in this respect.
4. The results deemed positive or negative must be reported in objective, quantifiable terms, such as, being determined by blood tests, scans, other techniques and instruments.

In spite of the "gold standard" status of RCT in the very recent history of evidence-based medicine, it is well to bear the following points in mind:

a. Not every treatment requires the *imprimatur* of RCT – in cases where the effects, say, of a drug are so obvious in immediately lowering mortality, no RCT is required. This was the case with sulphonamide and penicillin.[13]

b. Not every new treatment could/would be subjected to RCT on practical grounds, as there are far too many such treatments, continuously introduced into the market, by pharmaceutical companies, which differ from extant products, seemingly, in small ways.[14]

c. One should distinguish between two different contexts in which RCT are mounted – (i) by pharmaceutical industries, (ii) by the rest of the medical community which uses pharmaceutical products in treating patients.

We shall only comment upon the third point above, to show that the main aim of conducting RCT in the two contexts is different. Pharmaceutical companies are primarily interested in getting past the licensing legislation as laid out, for instance, in the US Pure Food and Drug Act or their equivalents in other, by and large, developed economies. Consequently, their remit is somewhat limited – all that RCT needs to demonstrate is that the product can perform in the way the label claims it does. For instance, a drug ultimately to be marketed (and patented) to treat depression must be backed by some evidence that it does have some effect in alleviating depression. It is not part of the remit of such trials to ensure how effective in reality it is compared to other drugs, or how safe it is in the longer run. It is unfortunate that the public and, indeed, even health professionals, misleadingly, believe that the licensing authorities are concerned, in particular, with the health and safety of the public.[15]

As we have already mentioned in Chapter 8, the true effects of a new drug cannot be ascertained by RCT conducted by pharmaceutical companies, as these trials necessarily have a short time limit; assessing the true impact from the health and safety point of view comes from long-term monitoring of the effects of a new drug when released for general use. This was indeed how the very serious side effects of the drug thalidomide were ultimately discovered. In the light of that catastrophe, guidelines for drug testing/use have been tightened but, when all is said and done, the ultimate "guinea pigs" are the patients in the world at large who take the drug which their doctors prescribe for them. Strategies for long term monitoring have been devised, such as the yellow card scheme – for instance, the scheme in the UK is operated by the Medicines and Healthcare Products Regulatory Agency (MHRA) which encourages not only health professionals but also members of the public to report to the Agency any side effects (adverse drug reactions) of a drug or vaccine so that it can identify previously unidentified safety issues.[16]

There are two main justifications for the use of RCT in the wider context of medical practice in society at large, which appear to be closely related: (a) whether the latest new treatment promoted by the pharmaceutical industry is more or less efficacious than the extant standard treatment(s); (b) as the latest product is often more expensive than the older options (especially in cases where patent rights have lapsed), is it cost-effective to endorse prescription of the former when money is always (relatively) in short supply? It is in this spirit[17] that the National Institute for Health and Clinical Excellence (Nice)[18] was set up in the UK in April 1999 whose main remit is to draw up guidelines about the use of a range of drugs.

We next turn to the main emphasis of this discussion about RCT, and that is, its relationship to the notion of controllability. We shall show that there is need to distinguish between two senses of the term "control" – (a) as shown above when RCT designs an experimental group: the group which would receive N, the new treatment, and group(s) which would receive S, the standard treatment(s). It is in this sense that assignment to such groups is methodologically required to be randomized. (b) The second sense which would be set out below has nothing to do with experimental and control groups in the first sense. It has to do with controllability/manipulation as a criterion of what constitutes "the cause", or "major cause" of a phenomenon in general, or of a disease in particular.

We have argued in the section above that the criterion of controllability is what is heavily relied upon in singling out one of the "inus" factors as "the cause". However, before proceeding further, let us look at another aspect of the history of RCT. Some writers on the subject have pointed out that not many realize that it was initially pioneered within epidemiology,[19] rather than clinical (bedside) medicine – the difference between the two, for the purpose here, may be construed as the study of disease patterns in populations, on the one hand, and the pre-occupation with individual patients succumbing to particular diseases, on the other. Lind's study of scurvy using a primitive prototype of RCT could be seen as an exercise in epidemiology. So, too, could Semmelweiss' study of puerperal fever. A famous nineteenth century example is Snow's study of cholera in London.[20] (John Snow is regarded as the founding father of epidemiology.) At that time, cholera epidemics frequently occurred in London; during the epidemic of 1854, Snow made a thorough study of the pattern of the disease in an area of the city which was supplied with water from two different water companies – one was the Southwark and Vauxhall Company, the other the rival Lambeth Company. Snow was struck by the following facts: (a) there was a pump in Broad Street supplied by the first company where the density of residents succumbing to cholera was high; (b) he also noticed that those working at a pub nearby did not succumb as the pub had its own water supply and did not draw water from the Broad Street pump; (c) he found that there were 71 deaths per 10,000 houses in this neighbourhood whose water was supplied by the Southwark and Vauxhall Company, whereas there were five deaths per 10,000 houses in the neighbourhood whose water was supplied by the Lambeth Company. (The mortality figure in the first case was 14 times greater than that in the second case; this was the crucial statistic rather than the gross figures of infection.) There appeared to be no other significant differences between the people living in the two neighbourhoods, save in one respect, namely, the water they drank and used supplied by the respective water companies. From this and other data, Snow concluded that the water provided by the Southwark and Vauxhall Company must have been contaminated. He ordered the removal of the handle in the Broad Street pump to prevent people from getting their water from the pump – the death rate fell dramatically. Snow had no idea about the exact nature of the contagious agent (as the discovery of the *vibrio cholerae* was not made till 1884 by Robert Koch), but he surmised it must be to do with some polluting matter from the sewerage which had entered the water supply somewhere along the chain.

We need next to say something about the notion of experiment in the context of controllability. Put simplistically, experiment is the foil of observation. The latter involves looking carefully at a situation and collecting data about it, to see what pattern emerges – it is what Bacon is commonly reported to have said about scientific methodology, and what the Humean analysis of cause set out briefly in parts of earlier chapters, is based on. As we have already seen, these accounts are not exactly convincing from either the philosophical or the scientific point of view. In contrast, experiment is not about the mere observation of the phenomenon under study; it is an active intervention upon the scene by the scientist/researcher. However, before proceeding further, one must first distinguish between three contexts in which the term may be invoked: (a) controlled experiment, (b) natural experiment, (c) field experiment. The discussion here is about the first, the strongest sense of experiment.[21]

Snow's handling of the 1854 London cholera epidemic illustrates well the distinction between observation and controlled experiment. First, he made observations (such as that people of the pub near the Broad Street pump did not succumb to cholera, as the pub had its own water supply, and so did not use the pump); he collected data (the gross figures of mortality in the two neighbourhoods, whose water was supplied by two different companies, as well as their differential rates of mortality), and so on. From these data, he surmised that the source of the illness could be the water supply, that the water could contain some harmful contaminant, and that the illness was probably contagious. If he were correct in this surmise, then disabling the pump in Broad Street ought to make a difference to the outcome. All this speculation (call it a hypothesis, if one so wishes) led him to conduct an experiment, focussing on removing the handle of the pump and then seeing what would happen. The before and after situations would remain in all other ways the same save for one variable/factor, namely, the water supply was no longer available. If the incidence of the mortality from cholera were to diminish significantly in the neighbourhood supplied by the Southwark and Vauxhall Company after the removal of that supply, then his hypothesis would have been verified;[22] if the mortality rate were to remain unchanged, then his hypothesis would have been falsified. As it turned out, the rate impressively diminished.

Snow's controlled experiment differs from the sort mounted by a pharmaceutical company in a drug trial. There the patients in both the experimental and control group were specially recruited for the purpose, then randomly sorted out. In Snow's case, his control group did not exist simultaneously as his experimental group; they were one and the same group in all respects, save that of the water supplied by the pump in the control group and its non-availability in the experimental group, later in time.[23] However, he deliberately manipulated this particular variable in his experiment, in order to study its outcome.

This is precisely what J.S. Mill (1806–1873) prescribed in his *System of Logic*, 1843.[24] In Book III, Chapters 8–10, he set out five methods for determining the cause of a phenomenon – agreement, difference, difference and agreement, concomitant variations, residues.[25] Snow's experiment, as well as drug trials conducted today, appears to rely, in the main, on the method of difference. Mill put the method of difference as follows (his Victorian prose style sounds stilted to our ears):

If an instance in which the phenomenon under investigation occurs and an instance in which it does not occur, have every circumstance in common save one, that one occurring in the former; the circumstance in which the two instances differ is the effect, or the cause, or are indispensible part of the cause of the phenomenon.

Let A stand for the neighbourhood whose water is served by the Broad Street pump.
Let X stand for the high occurrence of cholera in A.
Let B stand for the same neighbourhood but now no longer served by water from the said pump as it has been rendered non-functioning.
Let Y stand for the significantly reduced occurrence of cholera in B.
As A and B "have every circumstance in common save one", namely, the functioning pump in A but the same pump rendered dysfunctional in B, then the water from the pump (A) must be the "the cause" of the phenomenon X and Y is "the effect" of B.

Note that the emphasis of the method of difference on "every circumstance in common save one" is strictly speaking false. One can always find more than one circumstance in which the instances differ. Snow's study, for example, does not take into account genetic differences between the individuals in the Broad Street pump neighbourhood. He simply assumed that these differences were not material, but of course, we now know they could be, in explaining why some people succumb to a disease and others do not.[26] The default axiom is just that a difference, z, is presumed not to be relevant and significant unless one knows or suspects otherwise. However, the very idea that instances can have "every circumstance in common save one" has its natural home in engineering, where products are designed to be homogeneous and as near identical as they can be. When engineers test their products by setting up an experiment, they can divide the same products into two groups, but just making sure that the experimental group differs from the control group only in one "circumstance". The Millean idea, therefore, sits well within the machine context; and we have seen how the ontological *volte-face* involved in the human-is-machine perspective would render it appropriate for mounting tests in the medical context.

Mill claimed his methods to be canons of discovery and explanation. He can be criticized for holding both of these two claims. Take the method of agreement to illustrate its weakness as a canon of discovery. It says:

If two or more instances of the phenomenon have only one circumstance in common, the circumstance in which alone all the instances agree is the cause (or effect) of the given phenomenon.

Imagine the following: suppose you are trying to discover the cause of drunkenness by visiting a pub. Imagine, too, that some customers are male, others female; some are young, others middle-aged, yet others elderly; their clothes and hair style are all different; so are their accents, and so on. You observe that they have each ordered different volumes of their favourite drink, be this beer, wine, hard liquor. At the end of the evening, you note that all have become drunk. You then remind yourself that beer is very different from wine, and either is also very different from whisky or brandy; one ordered seven pints of beer, another four shots of liquor. What then is common to them all? Ah, beer, wine and liquor have one ingredient in common and that is water. Water then is surely the cause of drunkenness. You then announce your discovery. In this instance, following Mill's canon of agreement is no guarantee of discovering the cause of a phenomenon.

As for providing explanation of a phenomenon, we have already seen that Mill's methods are also not strictly-speaking relevant – from the scientific/theoretical point of view, it is Mackie's "inus" conditions or Collingwood's Sense III of cause which appear relevant. So, how best should we understand Mill's methods if they are neither about discovery nor explanation? We suggest it best to regard them in two related ways:

1. Singling out one of the "inus" conditions as "the cause" where cause is understood and defined in terms of controllability/manipulability. This is Collingwood's sense II of cause.
2. A Methodological guide as to how to set up an experiment via the singling out of a factor for manipulation in order to see if that factor satisfies the criterion of controllability – that is to say, whether by eliminating that factor, X (or introducing X into the test situation), it would make a difference. If, indeed, eliminating X leads to the disappearance of Y (or introducing X leads to the production of Y), then one may conclude that X causes Y. The meaning of this sense of cause is none other than that of controllability.

In Collingwood's sense III (or Mackie's "inus" conditions), not every one of the factors mentioned would be susceptible to controllability, as we may know no technique/technology by which we can eliminate it and then re-introduce it into a situation. In general, astronomical phenomena are not within human control; hence we cannot conduct controlled experiments regarding them. However, this is not to say that as technology advances and as we humankind feel that we should throw a lot of resources (intellectual, financial, material) at the phenomenon, we might not some day be able to mount such an experiment, or at least the simulation of one. The Large Hadron Collider (LHC)[27] underneath Geneva is one such example of the determination of the physics community, in particular, and of governments, in general, to get to the bottom, so to speak, of whether the Higgs boson[28] exists, to find out what could possibly have happened just after the Big Bang when the universe is said to have come into existence. By "recreating" the Big Bang, the experimenters are controlling a certain factor, the introduction of extremely high energy into the machine in order to determine its outcome.

In general, one can see that observation is not on the same epistemological plane as experiment – the former is relatively passive, the latter absolutely active. In principle, experimentation in the strict sense of the term is about controllability. If one cannot eliminate a factor, then re-introduce it or introduce it at will, one has not mounted a successful experiment. We know that conflagration would take place only in the presence of oxygen. But how do we "prove" or demonstrate that oxygen is a necessary condition for the occurrence of fire? Oxygen is present everywhere in the lower atmosphere surrounding Earth. Now if one could mount an experiment to remove the oxygen in the atmosphere, and see if a flame would continue to burn once the oxygen was eliminated; and, if the flame was extinguished, then we would have shown that oxygen is a necessary condition for conflagration. Today, a school laboratory could mount such an experiment should the teachers and their pupils be so minded.

Conclusion

This chapter has explored the following themes:

1. Clinical medicine is "practical" not theoretical medicine (in spite of the clear relationship between the two domains). As such, it is primarily interested in finding effective treatments either to cure a disease or to ameliorate its symptoms.
2. It is, therefore, concerned with a factor which is controllable or eliminable, making a difference to the therapeutic outcome. Such a factor is then called "the cause" or, sometimes, more cautiously "primary cause" of the disease.
3. This is Collingwood's Sense II of cause, where cause is constituted by the criterion of controllability/eliminability.

4. This is also the sense which underpins RCT.
5. The notion of control in RCT refers to two distinct though related matters: (a) "random control" is about designing two groups, the experimental group and the control group(s) using a random method of assigning patients either to N or S treatments; (b) control in this other sense is meant to single out a particular factor (one of the "inus" conditions), which the experimenter can eliminate from a situation at will (or then re-introduce into the situation) to determine the therapeutic outcome, as already mentioned.
6. Sense 5b above is about means/end rationality in terms of efficiency.
7. Senses 5a and 5b of control above are involved in the notion of experiment which is the foil to plain observation; the latter is primarily passive, the former characteristically active, as it is intended to be an intervention upon a given clinical situation, in order to make it tell us whether by manipulating a certain variable, a certain outcome would occur.
8. RCT presupposes that human individuals are uniform unless suspected otherwise, in which case such variables should be controlled. However, other things being equal, the patients in both the experimental and control groups are assumed to be uniform – this is its default axiom.

12

Epidemiology: "Cinderella" Status? What kind of science is it really?

Given the roles played by epidemiology in both practical (in saving lives, in diminishing suffering) as well as theoretical developments in biomedicine (such as its pioneering role in RCT) as set out in the last chapter, it seems odd that this discipline has (so far) not attracted the Nobel Prize in medicine. This chapter attempts to show that the Nobel Committee must have, in all probability, perceived it to be sub-standard science, and to explore the underlying reasons which could explain such a perception. Far from being the Cinderella of medicine, one can plausibly argue that it is "revolutionary science", not "normal science" in Kuhn's terms, once one grasps that it is a very different kind of science from atomistic science, that the metaphysics behind it – and in turn its entailing methodology as well as its notion of causation – are very different from the metaphysics behind the infectious-agent theory of disease and its entailing methodology as well as its notion of cause.

Is it revolutionary or sub-standard science?

In the last chapter, we referred to epidemiology as being concerned with disease patterns in populations. A slightly more expanded account may be found in Bhopal 2008, 3:

> epidemiology is the science and practice which describes and explains disease patterns in populations, and uses this knowledge to prevent and control disease and improve health.

We also referred in the last chapter to some famous cases in the history of epidemiological research. However, the most outstanding instance in the last six decades is the establishment of the link between smoking and lung cancer – as a result of this impressive series of research, governments in many parts of the world, from the 1990s, had begun to ban smoking in public space, as a public health measure.[1] This work began with the publication in 1950 in the *British Medical Journal* of an article ("Smoking and Carcinoma of the Lung") reporting the results of their research (a case control study) on the subject by Austin Bradford Hill and Richard Doll.[2] However, its findings, on the whole, did not convince the medical community that the link could be construed as a causal one. It took another study – this time, a cohort study[3] – involving 40,000 British doctors whose health was monitored for twenty years, to provide convincing evidence on this score.[4]

It is generally claimed and accepted that the work of Bradford Hill and Doll has transformed epidemiological research, putting it on an impeccable scientific footing from the methodological point of view; that Doll's substantial findings cover not only the tobacco/lung–carcinoma link, but also between other substances such as asbestos and cancer, radiation and leukaemia, alcohol and breast cancer as well as establishing that smoking increases the risk of heart disease. Their work in demonstrating that tobacco is a crucial factor in the production of lung cancer leading, as already observed, to the ban of smoking in public spaces and other measures to discourage smoking, has "probably prevented the premature deaths of millions already and ... may well prevent tens of millions more."[5] Doll is said also to be the most distinguished epidemiologist of the twentieth century.

Both Bradford Hill and Doll received knighthoods from their British sovereign for their contribution to cancer epidemiology. Doll was made fellow of the Royal Society in 1966, appointed to the Regius Chair of Medicine at the University of Oxford in 1969. International honours included the Presidential Award of the New York Academy of Sciences, an UN award, not to mention the King Faisal International Prize for Medicine (jointly with Richard Peto) in 2004. Yet, the Nobel Prize in medicine eluded him. It also eluded Bradford Hill (1897–1991) who died aged 93 years. Doll (1912–2005), too, reached the ripe old age of 92 years. Nobel Prizes cannot be bestowed posthumously – it seemed a sad fate for a man who, having lived so long to see his work on smoking and lung cancer finally accepted by the scientific community and governments world-wide, should, nevertheless, not be honoured by the world's acknowledged most prestigious prize in medicine. It almost seemed as if the Nobel Committee had deliberately avoided bestowing the honour on him and his co-researchers. Following his demise on 24 July 2005, the Karolinska Institutet in Stockholm announced on 5 October 2005 that the prize in medicine was to be awarded to Marshall and Warren for their discovery of *H. pylori* as "the cause" of peptic ulcer. The Nobel Committee had chosen presumably to honour the infectious-agent model of the monogenic conception of disease on the centenary of its award in 1905 to Robert Koch for his discovery of the tubercle bacillus as "the cause" of tuberculosis.

As the Nobel Committee practises a code of silence just as the jury system does about its decisions, there is no way by which one can find out what its reasons could be for awarding its prize to one type of research as opposed to another. However, one must assume that it has nothing to do with personal sentiments. Yet, what professional/scientific/theoretical reasons could there be for so pointedly (at least to all appearances) having ignored the achievements of Doll and his co-workers? For the sake of unravelling some of the presuppositions which could possibly be behind its decision, let us do a quick comparison between Doll (D) on the one hand and Marshall and Warren (M/W) on the other in certain aspects which could be considered to be relevant to the Nobel Committee's decision-making:

1. In terms of saving lives, both produce admirable results; however, as the tobacco smoking habit could in principle be acquired by a majority of the world's population while harbouring *H. pylori* (PUD) and actually manifesting peptic ulcer (PUD) is confined to a much smaller proportion of the world's population, D's impact in helping people to kick the smoking habit would save a considerably larger number of people from premature and painful death than M/W in terms of relieving the suffering of active PUD patients world-wide.

2. In terms of "paradigm shifts", although some people may argue that both warrant that label, one could, however, point out that W/M is not a real paradigm shift if that term is understood in the strict sense. It involves a shift only in a limited restrictive sense of the term, as in reality it is a piece of research which falls into what Thomas Kuhn calls "normal", not "revolutionary" science.[6] It is true that as a result of W/M, some peptic ulcer cases, today (at least where *H. pylori* [PUD which are also *PUD*] can be found) are no longer regarded as a physiological disease but an infectious one – in the past, the cause of all cases of peptic ulcers (PUD) was considered to be the mere excess of acid produced in the stomach. Apart from such a shift, W/M relies, by and large, on Koch's postulates, as we have seen – on the methodological front, unlike Koch in the late nineteenth century, M/W introduced no new methodological rules of procedure nor any new

forms of instrumentation to check their results. So, only in a generous interpretation of the term "paradigm" could W/M be said to amount to a paradigm shift, a kind of side-ways shift, as diagnosis as well as therapy have altered. However, this is not necessarily the way in which the term was most significantly used when Kuhn introduced it in *The Structure of Scientific Revolution*, 1962.[7] Examples given in the history of science of such truly radical shifts would typically include the change in astronomy from the Ptolemaic to the Copernican account, from Newtonian classical to Einsteinian relativity mechanics, from atomic to sub-atomic (quantum) physics, from Mendelian classical genetics to DNA/molecular genetics, and so on. The change from peptic ulcer (PUD) being understood as a physiological to an infectious disease (*PUD*), innovative though it may be, could not be said to be comparable to those just mentioned.

3. What about D? If the accolade normally given to D is a serious one, then D must be acknowledged to have put epidemiology on a scientific footing.[8] In other words, D would have transformed the subject from a "sub-scientific" status to one upon which the scientific *imprimatur* has now unreservedly been bestowed. In other words, in terms of scientific methodology, the subject should now be "pukka". In the history of science, it could be argued that D has initiated a paradigm shift comparable to that when the traditional craft-based techniques of breeding plants and animals over the millennia were put on a scientific footing with the re-discovery of Mendelian genetics right at the beginning of the twentieth century, to be followed by the new technology generated by Mendel's theory of inheritance a little over three decades later.

For a paradigm shift in this sense of the term to occur, there must be a consensus on instances of scientific research deemed to be exemplary. D clearly satisfies this fundamental requirement; the series of studies begun in 1954 on the link between tobacco smoking and lung cancer whose details have already been mentioned attest to it. If this interpretation of D is defensible and plausible, then it seems reasonable to argue that D is a much greater scientific achievement than M/W – D would amount to "revolutionary science" rather than "normal science", the marking of the birth of a mature science in a certain discipline. In other words, if M/W deserves the Nobel Prize in medicine, then surely D should also merit one.

Yet it appears not too strong to say that epidemiology remains the "Cinderella" of biomedicine. Not only did the most eminent epidemiologist of the twentieth century fail to be given the Nobel laurels, the subject itself is regarded to be marginal to medical training – in the main, it is not part of the undergraduate teaching curriculum as it is usually offered only as part of postgraduate medical training.

What then could really account for its "Cinderella" status? The section which follows will explore some of the presuppositions of "orthodox" biomedical research which appear to entertain an unfavourable opinion against epidemiology. We shall concentrate on two related matters in terms of (a) multi-factorial causation and (b) non-linear causation. We shall argue that such a notion of cause based on these two features is integral to epidemiology, a notion which is alien to clinical medicine, especially in its research methodology deploying RCT. It is this difference which may be responsible for the still continuing resistance to the reception of epidemiology as a proper discipline in biomedicine.

Causation: multi-factorial

The last three chapters (9, 10, 11) have set out in some detail the notion of cause as used in the monogenic conception of disease, in its use of the controllability criterion as well as of RCT. We have also said something about the linear Humean idea of cause. In other words, cause is understood to be mono-factorial and linear, and therefore, is in stark contrast to the causal stance adopted by epidemiology.

We have established that cause in an explanatory context is rarely mono-factorial – reality is multi-factorial. For the purpose of epidemiology, it is particularly concerned with three major types of variables and their interplay. These are: host, agent and environment, which constitute the so-called "triangle of causation".[9] Draw a triangle with the Host variable at the top, the Agent variable at the right-hand base and the Environment variable at the left-hand base of the figure. The Host is short for the person exposed to the agent in a particular environment who may or may not[10] develop the disease under study – the circumstances determining this would include age, general health, genetic factors, and so on. The Agent in question, which may be an infectious agent, a carcinogenic agent, would include the degree of virulence of such an agent, and so on. The Environment may include poverty, malnutrition, water contamination, air composition, workplace hygiene, overcrowding in the home, even seasons and the weather, and so on. One can immediately see that each of the three main types of variables covers many other sub-variables. It is the causal relationships between them all which ultimately determine whether the individual would succumb to a disease, and in how serious a form.

In the last few chapters, we have seen how the monogenic conception of disease which is mono-factorial deals with "the cause" of *PUD*. Let us here apply the epidemiological account of cause to see how it might accommodate the various data concerning PUD which the monogenic conception of disease ignores as they are deemed to be irrelevant in its construction of *PUD*. Some of these seemingly "anomalous" data include the following:[11]

a. 80 per cent of the world's population harbour *H. pylori*, yet 90 per cent of those with such a pathogen do not suffer from *PUD*, and hence, in the main, such individuals would remain ignorant that, as a matter of fact, they harbour the pathogen and, therefore, have *PUD*.

b. In developing countries, 80 per cent of the population may carry the bacteria. The incidence of infection increases with poor living conditions.

c. In developed countries, some persons with PUD do not harbour *H. pylori*; they are primarily the users of NSAIDs and aspirin.

d. Smoking, excessive drinking, extreme emotional/physical stress, are factors associated with PUD as well as *PUD* sufferers.

e. Sex as well as age appear to be implicating factors. In the past, peptic ulcer sufferers were predominantly male; now the ratio between the two sexes is 1:1. Peptic ulcers also increase with age – duodenal ulcers are found more frequently in the 35–50 age range while gastric ulcers occur more in the 50–70 age range.

On the epidemiology model of the triangle of causation, the data above fall under:

Host would include (a) – the occurrence or non-occurrence of symptoms amongst those who harbour *H. pylori*; (e) – age and sex; (d) – excessive drinking, smoking, extreme stress, physical/emotional.

Agent would include (c) – *H. pylori* (with *PUD)* or NSAIDs (without *PUD* but with PUD).

Environment would include (b) – poverty, malnutrition, overcrowding, air atmosphere, the weather, workplace hygiene and so on, which can affect the health of the individuals in the population with *PUD* including their immune systems, and therefore, susceptibility to PUD.

In the case of smoking and lung cancer, an analysis could be worked out as follows:

Host would include the smoking habits of the individual, their genetic inheritance, state of their health, their nutritional status.
Agent would include the carcinogenic nature of the many chemicals found in tobacco smoke, of which nicotine is only one.
Environment would include whether the space in which the individuals dwell consists of smokers, even if they themselves do not smoke, whether the space is enclosed or not, and so on.

The final outcome would depend on the complex interplay between all the sub-variables under the three main variables of the triangle of causation model. To emphasize such complex interplay, epidemiology also uses the wheel model of causation. For instance, the case of phenylketonuria handily illustrates the causal relationship between the gene/host relationship when that relationship operates within a larger environment which itself can be broken up into three sub-environments, namely, the physical, the social as well as the chemical and biological – for an image of this model, see figure 5.7 in Bhopal 2008, 135. The host possesses the gene defect which leads to an enzyme deficiency which in turn leads to brain damage. The host operates within a physical environment which includes facilities for early diagnosis, as early diagnosis backed up by dietary control could help prevent/control the manifestation of the disease. The host also operates within a social environment which includes active support to make it possible to sustain on a long term basis the special dietary regimen, if prevention/control were to be effective. Such special diet, if conscientiously sustained, would mean that the host's chemical and biological environment would be such that the disease would not manifest itself in the lifetime of the host. This enlarged wheel model of causation shows very clearly that it would be distorting to regard phenylketonuria as a genetic disease (that its cause is a gene defect) *tout court*; it is also an environmental disease.

In other words, according to epidemiology, a disease does not have a single cause. This is a singularly crucial difference between it and the monogenic conception of disease. This difference would certainly account for why those wedded to the latter would have such difficulty in accepting the former as "proper" medical science. For epidemiology, it is a mistake to single out one of the numerous variables as "the cause" of a disease from the scientific/explanatory point of view, which is what the monogenic conception of disease enjoins one to do. All the relevant variables constitute a set of "inus" conditions. However, as we have seen in the last two chapters, should one wish to single out one of these as "the cause", its selection is made within a practical/clinical context using the criterion of controllability. Diet as well as smoking are in principle controllable and are within the effective control of the individual, given support by the community/society. This, as we have earlier argued, is Collingwood's Sense II of cause in contrast with his Sense III which is the sense used in the scientific/explanatory context.

Non-linear causation and post-postmodern ecosystemic science

We next have to say something about its non-linear character. In Chapters 9 and 10, we have talked in outline about Humean causation which is linear. However, before going further, we must enter a caveat – Bhopal 2008 could be said to be a leading text in epidemiology today, but the author does not use the term "non-linear causation", although what he writes appears compatible with it or implies it. Hence, what follows may or may not be a misunderstanding of Bhopal; this author, all the same, would like to argue that while, on the one hand, epidemiology and non-linear causation go together, linear causation goes with the monogenic conception of disease, on the other.

Let us immediately give a quick summary of the major points of difference between the two conceptions of causation:

Linear	Non-linear
Humean	Non-Humean
Mono-factorial	Multi-factorial
One cause, one effect	Inter-acting causal variables, one effect
Causal direction moves in straight line	Causal is reciprocal from A to B, B to A...
Static, ahistorical	Dynamic, historical
Negative feedbacks	Positive feedbacks – new equilibrium
Reductionist methodology	Non-reductionist methodology
Atomistic materialism	Ecosystemic/holistic materialism

Given the differentia set out above, perhaps the easiest way to understand non-linear causation in epidemiology is to see epidemiology as an "ecosystemic science" as opposed to the "atomistic science" established in the seventeenth century. The latter is what we normally call "modern (Western) science" whose characteristics have already been discussed in detail in Part I. The former has emerged more recently, when ecology became accepted as a science in the way understood today after World War II, especially through the work of Eugene Odum.[12]

We have seen in Part I that modern science was/is informed by empiricism-cum-positivism. In the last few decades, the term "post-modern" has appeared which rejects the certainty of that philosophy, thereby embracing relativism in one form or another as philosophy.[13] Ecology and the sciences inspired by it are neither modern nor post-modern but may be called "post-postmodern", as they reject the more simplistic account of modern science (which gives rise to its claim to certainty) as well the philosophical drawbacks of relativism; instead, they operate within a much more complex framework of cause and effect.

The term "ecosystematicity", though infelicitous, may nevertheless be apt, as studies of ecosystems in ecology have amply demonstrated the relevance of the concept of non-linearity as causal reciprocity. Any ecosystem in real life involves numerous agents.[14] These are often too numerous to be exhaustively identified as individuals, so for the sake of simplicity, we resort to enumerating them in terms of classes or groups. Three main groups may be identified: living agents which are human; living agents which are non-human; non-living agents.[15] Theorists disagree about the composition of ecosystems on our planet. Some hold that without exception they involve all three types of causal agents; others (a minority)[16] hold that although it is getting more difficult to find ecosystems which can be said to be free from human intervention, it, nevertheless, makes sense to talk about ecosystems in the absence of human presence and manipulation. However, all theorists agree that all ecosystems (on Earth) include non-human living as well as non-living agents. In a study of any ecosystem, the crux of the matter lies in the complex causal relationships which occur between the biotic (in whatever form) and the abiotic. This holds true at any level of investigation, whether the ecosystem is micro or macro – a handful of soil is as much an ecosystem as a virgin forest which may be the size of England and Wales combined.

The causal relationships between the interacting agents involved are reciprocal. Take the following: a hair-line crack exists in a rock (A); water (B) enters the crack, turning into ice, thereby enlarging the crack in the process; a seed floats by, lodges itself in the crack and grows (C); C together with B cause A to widen, which in turn permit more water/

ice to enter, giving more space for C to grow by widening the crack still further with more rain/frost (B) at the same time entering and eroding it, and so on.

As another instance of "ecosystem science", global climate change[17] naturally involves many, infinitely more numerous causal factors than the example just cited. The abiotic, here, minimally includes Earth's atmosphere; the emission of CO_2, and other greenhouse gases; oceans; particulates/aerosols in the atmosphere, clouds, water vapour. The biotic minimally includes humans who release CO_2 through burning fossil fuels; cattle and humans releasing CO_2 (when they breathe) and methane (when they fart and defecate, and when humans grow paddy rice); young trees absorbing CO_2 from the atmosphere, decaying or felled trees releasing the gas. Large-scale deforestation without adequate replanting upsets the balance between the uptake of CO_2 and its release.[18]

As the "reciprocal"[19] or "ecosystemic" concept of cause is much less familiar, we must elaborate a little more.[20] This schematic model – with two agents, for the sake of simplicity – of A producing an effect on B, B in turn producing an effect on A could be found in nearly every domain of investigation, whether it concerns phenomena in the physical world which constitute the subject matter of the natural sciences, such as ecology, geology, biology/genetics, or whether it concerns phenomena in the social world which constitute the subject matter of sciences such as economics, psychology. The following illustrations are taken from the latter. Imagine parents/teachers telling a child that he is stupid, pointing to his poor results at school (A1). The child reacts to A1 by tacitly agreeing with it, doing nothing as a result to improve his performance (B1). B1 in turn produces even poorer results (A2), which could lead to further deterioration in the child's self-image and hence performance (B2). This process of reciprocal reactions in psychology is sometimes called 'self-fulfilling prophecy' – give a person a bad name and hang him. In economic behaviour, it could lead to a run on a bank – start a rumour that a certain bank is about to go bankrupt; if a sufficiently large number of its clients believe it to be true, they would immediately withdraw money from the bank, thereby rendering it bankrupt, even though its bankruptcy began with a mere rumour.

A simple example in clinical medicine would be the following: at t_1, a person (A) finds himself itching in a particular part of the body; at t_2, to ease the itch (B), A scratches B; at t_3, A's scratching, far from easing the itch, increases it as scratching makes the skin react with greater ferocity than in the absence of such intervention; at t_4, when the itch intensifies, A resorts to scratching it even harder, and so on, until the skin becomes so raw and so bloody that at t_5, A could no longer even scratch B.

Thus, a new level of equilibrium has been reached. This shows not only the reciprocal nature between two events, playing in turn cause or effect, but also the dynamic, historical nature of the causal relationships between the events/processes involved.

Epidemiology tacitly uses this understanding of cause. The epidemiological understanding of phenylketonuria may again be invoked to illustrate it. Whether the disease will manifest itself in an individual will depend on:

1. At t_1 (just after birth), the individual is diagnosed with the genetic defect.
2. At t_2 (immediately after diagnosis), an appropriate diet must be prescribed and adhered to.
3. At t_n, that is, throughout the lifetime of the individual (provided the diet is conscientiously adhered to), s/he does not manifest the disease.

These show that the relationship between the events mentioned in them are historical, dynamic and reciprocal – the gene and its expression depends on an intricate relationship between it and its environments, physical, social, biological/chemical. This is also to say that the epidemiological understanding of phenylketonuria is that its causation is multi-factorial, and cannot therefore be simplistically defined as a genetic disease, as it would be according to the monogenic conception of disease.

Epidemiology, like ecology, is concerned with community, populations both on a local as well as a global scale. Snow's nineteenth century study of cholera was confined to a particular part of London; Doll's and his co-workers worked on larger scales.

Unlike clinical medicine, epidemiology is not so much interested in the fate of individuals as patients but more in preventing the emergence of disease patterns amongst communities and populations and/or improving the health of such communities and populations. Snow was keen to work out why one neighbourhood should suffer a cholera mortality rate 14 times greater than another neighbourhood, rather than study why this particular individual died of cholera. However, one should not misunderstand this to mean that successful epidemiological research would have no impact on individuals, as it clearly would have – once the handle of the pump in Broad Street was removed, the death rate fell. This meant that some individuals in the community who would have died, lived instead. While clinical medicine concentrated

on identifying the infectious agent and producing an effective form of
treatment against the disease, epidemiology concentrated on a public
health measure to prevent a certain disease pattern from emerging.

This leads us into exploring the respective metaphysical frameworks
in which the monogenic and the epidemiological conceptions of disease
are embedded. To illustrate the difference, take vaccination as immunization against infectious diseases, a typical epidemiological strategy.
Such a programme is predicated upon the following:

1. It does not require 100 percent co-operation but a significant proportion of the population must be vaccinated for it to succeed. This is
   called herd immunity.
2. Once there is herd immunity, an individual may take advantage of
   its existence to avoid vaccination for himself/herself (or for his/her
   child). Furthermore, if the risk of a specific vaccination is perceived
   to be high and/or serious, then the individual may feel strongly
   tempted to "free ride."
3. Yet if sufficient individuals were to "free ride", then the success of the
   programme would be undermined.[21]

Vaccination programmes may be distinguished from general screening
programmes for non-infectious diseases – while the former belongs to
epidemiology proper, the latter may readily be seen as part of monogenic medicine. Cancer is not an infectious illness; hence the success of
a screening programme for breast cancer does not depend on its uptake
in causal terms. It would be nice if all women at risk were to come forward to take the test but causally speaking, the benefit is individual
and has no fall-out for the community at large; there is no equivalent
to herd immunity in this kind of programme which is, of course, also a
part of preventive medicine.

The herd immunity phenomenon illustrates very well the notion of
population thinking in epidemiology. As "ecosystemic science", epidemiology must understand cause not simply in individualist terms. We
have seen in Chapter 10, cause as understood within "atomistic science"
is additive in nature; each straw put on the camel's back – in terms of
its causal impact – is identical. But when cause is viewed in ecosystemic
terms, the effect of each act of vaccination is not identical. When the
uptake of vaccination on the part of individuals within a population
reaches a certain critical mass, then herd immunity emerges as a phenomenon. When such a stage is reached, avoidance of vaccination on
the part of a certain limited number of individuals is of no significance

in causal terms (although it may have significance in moral terms, as we have earlier commented).

In other words, epidemiology is best understood with reference to holism as its metaphysics. In Chapter 4, we have already shown how a whole can have properties which cannot be reduced to the sum of the properties of its constituent parts. In ecosystem thinking, the same metaphysics obtains, engendering a non-reductionist methodology. Take, for instance, a patch of soil as an instance of an ecosystem. In the soil, you find individual constituent parts, such as very small as well as larger pieces of stone, tiny gritty bits of rock so small that they may not be visible except under a microscope, some dead roots and remains of plants, micro-organisms, moisture/water, air pockets between these parts, and so on. Each of these constituent parts taken on their own is not soil; their complex causal relations with one another add up to soil. The soil has emergent properties, such as its own peculiar texture, its own smell even, which cannot be accounted for solely in terms of the properties of its constituent parts.

Furthermore, in ecology as well as in epidemiology, there is a nesting of ecosystems – one ecosystem is nested within a larger one, this in turn within a still larger one, until one reaches the largest ecosystem on Earth, which is the atmosphere, with its specific composition of different gases varying at different sea levels, the wind, clouds, mists, frost, snow, dew, sunshine, and so on. The shady corner in a meadow for certain purposes may be regarded as one small ecosystem, nesting within a larger ecosystem which is the meadow of which it is a part; the meadow in turn is part of a larger ecosystem, such as the hills and stream surrounding it, which in turn is part of some bigger ecosystem.... In epidemiology, the wheel model of causation shows, for instance, how the genetic inheritance of an individual may be deemed to be an ecosystem which nests within a larger ecosystem which is the body of the human individual with its own physiological, biochemical systems; this body in turn nests within a larger social/cultural ecosystem called community/society, within a still larger physical ecosystem which ultimately includes the Earth's atmosphere. This perspective accounts for the fact that the causal relationship between a gene and its expression is an immensely complex matter;[22] hence, no disease can just simply be called a genetic disease in the majority of cases.

It seems plausible to conclude that epidemiology and the monogenic conception of disease are respectively embedded in different metaphysical and therefore also different methodological frameworks. The latter belongs to atomistic science which also historically goes with the body-is-machine ontology; the former is part of what is called here "ecosystemic science" which is holist in ontology and non-reductionist in methodology. It also follows that they deploy different notions of cause in their explanations of disease. All these differences may account for why epidemiology is not deemed deserving of the Nobel Prize in medicine, as the Nobel decision-makers and their advisers may be too wedded to the monogenic conception of disease, and hence fail to appreciate the paradigm shift which epidemiology has brought about.

Epidemiology and controllability

In the last chapter, we showed how both clinical medicine and epidemiology use the criterion of controllability in the practical/therapeutic context; they both mount experiments. However, we also commented on some crucial differences between them, such as that randomization and double-blinding fail to obtain in epidemiological experiments. In this section, we shall further explore some of these differences, which again, whether consciously or sub-consciously, might have influenced the Nobel Committee not to look favourably upon epidemiology.

One must grasp that epidemiology is primarily preventive medicine, whereas clinical medicine is, by and large, therapeutic medicine.[23] The former bestows benefits on humanity through measures, if implemented in the public domain, which would ensure that an "ill", such as the frequency of a disease occurring, would greatly diminish, and correspondingly "a good" such as better health would ensue. The latter bestows benefits on humanity by removing or ameliorating a disease affliction which has, in the main, already befallen on certain individuals, and in this sense ensures that better health would ensue. Drugs are typically used in the latter, while other strategies/techniques such as securing a clean and safe water supply in particular, a more wholesome, non-polluting environment in general, are typically used in the former.

However, one obvious reason why epidemiology, unlike clinical medicine, cannot mount its experiments as RCT is not methodological but ethical. We, collectively as society, are of the view that research must not deliberately and intentionally inflict pain and suffering on the subjects involved in a scientific experiment. Take the research done on the link between smoking and lung cancer. To test the research finding further to ensure that a direct major causal link exists between these two kinds of phenomena, there is, in principle, no difficulty, in mounting RCT. One could recruit a certain large number of people (even several hundreds of thousands, if necessary, world-wide), all non-smokers, equally matched for age, sex, state of general health, nutritional status, occupational status (for instance, professional/ middle class workers not exposed to obvious forms of environmental pollution at work), and so on; randomly assign them to an experimental group and a control group.[24] Over a period of 20 years, ensure that the experimental but not the control group smoke, say, 25 (or 40) cigarettes of similar strength and design daily, while ensuring that those in the control group do not. At the end of the experiment, researchers would compare the rate of lung cancer in the experimental group with that in the control group. If the difference is statistically significant, then one could conclude with confidence that tobacco smoking is indeed a primary cause (if not the only cause) of lung cancer. However, ethical constraints prevent such a kind of research to be carried out.[25] Society's ethical outlook even in the past, never mind today, would not dare openly endorse it,[26] as smoking tobacco is predicated upon the belief that it is harmful to humans.

In contrast, in a clinical trial, the research is predicated upon the belief or hypothesis that the new drug (or treatment) under test would bring benefit to the participants but not on the assumption that it would definitely bring less or no benefit. (Of course, it may turn out to bring less benefit than an existing drug for the same condition or even bring positive harm; but whether it is does or not is indeed the point of the exercise.) In such a context, one is ethically permitted to conduct RCT.

In the last chapter, when we looked at Snow's work on cholera in the nineteenth century, what Snow did not and could not do was to mount RCT (even if the RCT concept as understood today had occurred to him). Apart from randomizing and double-blinding, RCT also requires that the experimental and control groups exist simultaneously in time. Snow could not randomize, double-blind nor ensure that the two groups existed in the same temporal frame. Instead, he simply "made do" with an arrangement which fell far short of the gold standard of RCT. Snow's control group existed prior in time to the experimental group – the former consisted of the neighbourhood which used water from the Broad Street pump and the latter of the same neighbourhood which no longer used the Broad Street pump as its handle was removed. As the participants of the two groups remained identical, in that sense, Snow may be said to have equally matched participants in all aspects save one. In the latter half of the last century Doll and his co-workers have devised other methodological designs, apart from Snow's kind of experiment or trial. However, they are all perceived to fall short of the

Platonic form, so to speak, of methodological perfection found in RCT; they are judged to have failed to meet the gold standard requirements. Bhopal (2008, 15) has conveniently set out five epidemiological designs in an ascending order of "methodological goodness" which are ultimately concerned with forming hypotheses to test the possible causes of diseases:

a. Case series and population case series, whose main aim is to determine the rates of target cases to population data, to analyse the resulting patterns.
b. Cross-sectional which studies a population within a defined time and space to ascertain the proportion of those afflicted with a specific disease and those not thus afflicted.
c. Case-control which by setting up a series of cases against a control group enables it to look for similarities as well as differences between the two.
d. Cohort which monitors populations, relating their "medical fates", so to speak to risk factors to which they may be exposed.
e. Trial which involves an intervention designed to improve the health of the subjects, and then determines if the intended effect occurs.

It seems fair to conclude that these methodological designs are the outcome of either ethical or practical constraints placed upon epidemiological research; as a result, they are perceived to be "flawed" when compared to the "pukka" RCT. This may constitute another reason why the Nobel decision-makers have not seen fit to bestow its accolade on epidemiological research.

Conclusion

1. In this chapter, we have tried (a) to argue that epidemiology is perceived by the Nobel Committee to play "Cinderella" to the monogenic conception of disease and its close relative, clinical medicine, (b) to set out some reasons which might have led to such a perception.
2. In the fairy tale, Cinderella turned out to be the one who won the Prince's heart; she was misperceived by her step-mother and step-sisters to be unattractive, and treated as an "underdog". Analogously, could it be that epistemology misses out on the top accolade, called the Nobel Prize in medicine, because the orthodox establishment of which the Nobel Committee is a part has failed to realize that the work of Doll and others have lifted the subject of enquiry to a new

plane, that such research has produced a paradigm shift in medical thought and practice. In other words, the Nobel Committee does not perceive epidemiology to be a new "revolutionary science" in Kuhn's terms.

3. What sustains the perception may lie in the failure to realize that epidemiology operates within a different philosophical framework from that which underpins the infectious-agent conception of disease and research in clinical medicine, which belong to atomistic science and its reductionist methodology. Epistemology, on the other hand, belongs to ecosystemic science and its non-reductionist methodology. The former relies on linear mono-factorial causation, the latter on non-linear/reciprocal, multi-factorial causation. Both are forms of materialism – the former is a perfect exemplar of modern science dating from the seventeenth century, based on atomistic materialism, the latter of post-postmodern science of the late twentieth and twenty-first centuries, based on ecosystemic or holist materialism.[27]

4. Another perceived flaw could lie in the methodological designs pioneered by epidemiological research, borne out of ethical as well as practical constraints, which dictate that it cannot set up experiments and trials according to the specifications endorsed by RCT.

Conclusion

This book has explored the implications of the philosophy as well as its entailed methodology for science in which modern medicine (or biomedicine) is embedded. The main theses pursued are as follows:

1. Every science is set within its own philosophical framework, its own metaphysics (ontology), epistemology, and so on. In turn, its metaphysics-cum-epistemology entails its own methodology for doing science. Thus (Western) medieval science and its methodology operated within a philosophical space which was primarily Aristotelian; its scientific explanatory schema, therefore, was Aristotelian, in terms of the four causes – formal, final, material and efficient.

2. In changing from the medieval to the modern world-view in the seventeenth century in Western Europe, the most revolutionary axiom of the latter is the ontological *volte face* that the naturally-occurring world (including its organisms) is artefact, more specifically, a machine. As biomedicine is but a part of modern science, it follows that built into it is this momentous ontological *volte face*, according to which the (human) body is machine. In articulating this project, many leading philosophers contributed. In particular, Descartes introduced his mind-body dualism which allocated to biomedicine the human-body-is-machine as its domain of understanding and therapy.

3. It follows from the ontological perspective of the human-body-is-machine that MEDICINE is ENGINEERING and that medicine is a form of engineering. The ENGINEERING approach accounts for the ways in which both the theoretical biomedical sciences (such as anatomy) as well as medical technologies (such as pharmacology and surgery) are conducted and designed.

4. The ontological *volte face* therefore entails that the grand tradition in biomedicine is atomistic in metaphysics, reductionist in its methodology both in the theoretical and therapeutic domains; furthermore, it also invokes a notion of cause which is Humean, mono-factorial and linear. This cluster of features is characteristically displayed in one powerful aetiological definition of disease, namely, the infectious-agent model which is monogenic in conception.

5. The above conception of disease has reigned supreme for over a hundred years, ever since the germ theory began to bear impressive fruits towards the end of the nineteenth century. It remains a progressive research programme even today in spite of the fact that it has run into some serious anomalies. Two may be mentioned here: (a) recent research shows that the placebo effect is a near-universal phenomenon which obtains across all forms of medical treatment (orthodox or perceived to be quack by the orthodox medical establishment), (b) the promise of bespoke medicine via SNPs. These seem to challenge the very foundation of the dominant conception of medical disease – for instance, the placebo effect (as just set out) would appear to undermine the epiphenomenalism which serves as the philosophical foundation for psychopharmacological interventions; the existence of SNIPs appears to undermine the fundamental axiom of the repeatability of (medical) scientific experiments as well as its related methodological gold standard, RCT.

6. At the same time, the continuing ascendancy of this particular dominant tendency may serve to obscure the newly acknowledged fact that disease causation is very rarely mono-factorial. However, the focus on the monogenic conception of disease has also served to side-line the achievements of another tendency, co-existing with it since the nineteenth century. This is epidemiology. By the fourth quarter of the twentieth century, epidemiology has finally appeared to have achieved the status of science.[1] In contrast to its more successful operator, this medical science is holist in its metaphysics, non-reductionist in its methodology and its notion of cause is multi-factorial, non-linear or reciprocal. This may be called "eco-systemic" science. However, judged by the standards deployed by the monogenic conception of disease, this approach appears to have been judged "sub-standard". [2] Paradoxically, it may be said to be "revolutionary" science. Its achievements since the 1970s have been acknowledged to be impressive, especially that of establishing a convincing causal link between smoking and lung cancer. Yet no Nobel Prize has been awarded to its eminent practitioners; instead, in 2005,

on the hundredth anniversary of the prize given to Koch in 1905, the Nobel Committee chose to award its prize in medicine to two researchers who simply carried out the "normal" science established by Pasteur, Koch and others from the 1880s. The "normal" science of Koch, Warren and Marshall belong in philosophical/methodological orientation to the nineteenth century while epidemiology belongs, one may plausibly argue, to the last quarter of the twentieth century as well as to the unfolding twenty-first century which may be marked by the rise of "ecosystemic" sciences.

7. The monogenic conception of disease is, perhaps, best understood in the context of practical, clinical/therapeutic medicine rather than in the explanatory/scientific context of understanding the nature of diseases. It selects the infectious agent (or other single specific factor) as "the cause" of a disease with the expectation or hope that it can be controlled via a certain treatment, whether this is an antibiotic, a vaccine or genetic modification, and so on. The criterion of controllability/eliminability is constitutive of the notion of "the cause" in clinical medicine – it is what Collingwood has called Sense II of the term, in contrast to what he calls Sense III used in the explanatory/ theoretical context, and which can be said to refer to what Mackie calls "inus" conditions of causation.

8. The infectious-agent theory of disease on the one hand, and epidemiological research on the other, upon careful analytical scrutiny, bear out a main contention of this book, namely, that each conception of science/medicine is embedded within a particular philosophical framework and the methodology that such a framework entails.

Notes

Introduction

1. More accurately, this medicine should be referred to as "modern Western medicine" (but now globalized); but for brevity, this chapter and others to follow will simply use the term "modern medicine". (However, Chapter 7 will show that the term "biomedicine" is, all things considered, perhaps, the most appropriate term to use.) The term "Western" will also, in general, be omitted before "modern science" or "modern philosophy", even though, these subjects, too, originated in Western Europe in the seventeenth century.
2. As we shall see, this date, on the whole, is probably an oversimplified convenient starting point.
3. This insight is very rarely expressed or explored – for one exception see King 1978, 178 who has written:
   > Underlying the acceptance or rejection of a given explanation is the total world view, the basic metaphysical outlook that a person might hold. Some philosophers make this world view highly explicit and carefully articulated. More often the world view is part of the general intellectual environment, accepted without question.
4. For a detailed ontological exploration of the distinction between the natural and the artefactual, see Lee 1999.

1   Philosophical Foundations

1. *Novum Organum Scientiarum,* written in Latin and published in 1620.
2. This concession in the Anglo-Saxon tradition of scholarship is a grudging one – witness the attempt in UK universities in the 1980s, under political pressure from the government of the day, to change the name of the faculty of social sciences to the faculty of social studies.
3. Axiomatic systems consist of definitions of certain basic terms upon which certain fundamental axioms are erected. Take monotheistic religions, such as the Abrahamic ones. "Monotheism" is defined as a religion which postulates the existence of one divine entity. This entity in turn is defined in a certain way, namely, that it is omniscient, omnipotent as well as all benevolent/merciful. From these definitions and their entailed axioms, the theologian would be able to construct a systematic set of propositions about the divine deity and its relationship with the world. Analogously, Euclidean geometry defines its basic terms "point", "straight line" in certain ways; from them and their entailed axioms, (Euclidean) geometricians would be able to construct and deduce a systematic set of propositions which include the following famous axiom: from a point above a straight line, only one other straight line can be drawn which is parallel to that straight line.

4. We hope to demonstrate this to some extent in this and the next chapters.
5. However, this book is not primarily concerned with this kind of issue in applied ethics; its discussion of the relevance of values to medicine lies elsewhere – see Chapter 3.
6. There are philosophers who maintain that knowledge is no more than justified/warranted belief.
7. Of late, some logicians have started to challenge bi-valent logic. For instance, since 1965, one takes the form of multi-valent or "fuzzy logic" (associated with Lofti Zadeh), while another since the 1970s takes the form of paraconsistent logic (whose most prominent practitioner today is Graham Priest in Australia). For quick reference to the former, see http://plato.stanford.edu/entries/logic-fuzzy/ and for the latter, see http://plato.stanford.edu/entries/logic-paraconsistent/.
8. Descartes advocates in general that philosophical, including moral, notions are clear and distinct ideas.
9. One outstanding representative is a form of utilitarianism whose major champion is Jeremy Bentham. The influence of Bentham's utilitarianism can be found in cost–benefit analysis in general, and in the allocation of resources to medical services and the saving of lives in particular.
10. However, the doctrine itself is older than that and may be traced to its first systematic formulation by Thomas Hobbes in his famous book, *Leviathan*, published in 1651 – see Lee (1989a, chapter 2).
11. This is, however, not to imply that Aquinas was not open to the ideas of other philosophers such as Plato, but that he relied in the main on Aristotle, absorbing from other thinkers what he could reconcile with his reception of Aristotle's philosophy.
12. Aristotelianism did not operate in the world of Eastern Byzantine Christianity.
13. However, some scholars maintain that it should more accurately be called "scholasticism".
14. Comte himself was less harsh about it than his twentieth century successors, members of the Vienna Circle, who called themselves "Logical Positivists", as he was less reluctant to regard it as solely backward or sterile, conceding that the metaphysical conception of the world over several fields contributed significantly to the advancement of knowledge. Those who followed him were not so charitable.

2 Modern Philosophy, Modern Science and Its Methodology

1. Admittedly, Galileo, unlike other theorists (such as Hobbes or later Locke) neither found the time nor felt that the times were ripe to formulate a system of philosophy to match the method and findings of the new science and to challenge explicitly the old philosophy. Nevertheless, his method itself embodied many of the tenets which later made up the core of the new philosophy. Furthermore, he made the distinction between primary and secondary qualities of material objects (the former pertains to their shape,

size, motion which form the subject matter of physics, the latter to their colour, taste, smell, and so on, which can only be subjectively ascertained and hence stand outside scientific investigation), a distinction which Locke (1631–1704) later systematically incorporated into his version of empiricism in *An Essay Concerning Human Understanding* 1690.

2. Note that one ought to distinguish between two senses of "laws of nature" – (a) the first understands "laws" in terms of norms meant to hold universally across cultures and historical periods (such as, it is a law of nature that criminals ought to be punished, or that punishment ought to match the gravity of the crime); (b) the second has nothing to do with normative conduct but simply with the observed behaviour of natural phenomena which obtain universally across space and time (such as that night and day follow each other).

3. It rejects revelation (reference to scriptural/holy texts) as well as intuition. Regarding the latter, amongst the modern greats, only Descartes stands out as an exponent, advocating "clear and distinct ideas" which are self-evidently or intuitively true.

4. *A System of Logic* 1843.

5. See Chapter 9, Section entitled "Postulate 1 and the monogenic conception of disease": the flaw pointed out there is that inductive logic of a Humean kind cannot do justice to the concept of cause. Chapter 10, section entitled "Humean Roots" pursues the less trodden ground that the Humean account of cause is defective because it is mono-factorial and linear. However, the most influential critic of inductive logic in the last century was Karl Popper who held that not only is inductive logic fundamentally flawed, but that science does not use nor need it, and that deductive logic is the logic of science – see Popper 1959.

6. According to some critics, the symmetry thesis overlooks the fact that although some systematic study of natural phenomena may yield satisfactory explanation, nevertheless, it does not enable one to make satisfactory prediction about such phenomena. Does it follow that such studies do not count as science? If so, the term "science" may be too narrowly defined and is, therefore, in danger of excluding what to all intents and purposes count as "pukka" sciences, such as evolutionary biology or geology. The theory of evolution and its mechanism of natural selection can explain an impressive array of biological phenomena, such as the demise of a particular species or what new species would arise to occupy the habitat vacated by the declining one, yet it is not in a position to predict with any real precision (as astronomy can the next eclipse of the sun). At best, it can only give a rough characterisation of what might happen. Geology can explain why and how earthquakes happen but not predict when exactly the next earthquake in a certain region would occur; it can explain why oil is found in certain types of rock formation but is not able to predict where exactly the next successful oil well may be dug and found. Nor can it predict when a volcano may next erupt, although it can adequately explain why and how volcanic eruptions occur when certain conditions obtain. A rather ludicrous situation has happened recently (June 2010) following an earthquake in L'Aquila in the Abruzzo region of central Italy on the morning of 6 April 2009.

The main quake registered 6.3 on the Richter scale. All told, it left at least 290 dead, 1000 injured, 40,000 homeless, and 10,000 buildings damaged or destroyed – see http://www. earthmagazine. org/ earth/article/23a-7d9-7-e ; http://earthquakes.suite101.com/article.cfm/ laquila_ italy_ . A year or so later, some thirty citizens of the city have petitioned the city magistrate to bring the charge of manslaughter against the seismologists of the region on the grounds that they should have foreseen the onset of the quake based on small quakes already registered before the event. The public prosecutor seems not to have appreciated that while lots of indicators of impending earthquakes are on offer, none of them appears to have survived the test of success demanded by scientific methodology – see http://www.mi. ingv.it/open_letter/

3   Category *Volte-face*: Organisms for Machines

1. According to one interpretation of Chinese cosmology, it is an instance of the latter (although it is also correct to point out that Chinese cosmology before the Zhou dynasty was also understood in straight-forward shamanistic, divinatory terms) – modern European and traditional Chinese cosmologies are both forms of materialism. The crucial difference between them is that the former is atomistic (and, therefore, reductionist in its metaphysics-cum-methodology) whereas the latter is organismic and holist in its metaphysics-cum-methodology. For a succinct account of the former, see Merchant 1980; for a brief exploratory account of the latter, see Lee 2011.
2. Merchant 1980 is an excellent example of such a view; see also "Metaphors of Human Biology" in Temkin 1977.
3. Merchant may again be cited as an important scholar who holds such a view.
4. For further exploration of the difference between the two ontological modes of being, see Lee 1999.
5. On the history and philosophy of technology, see Lee 2005, Chapter 2.
6. This is an oversimplification, holding true only in the early phases of technological development. Since the 1840s, modern technology generated by discoveries in modern science can transform naturally-occurring material, such as oil into plastics. Today, plastic is a commonly used material but it does not occur in nature. For details, see Lee 2005, Chapter 2.
7. To put the point more carefully and in greater detail, one can say that there are three categories of organisms to be distinguished from the ontological perspective: (a) Those which are genuinely naturally-evolved, not the result of human manipulation and control. This is the group which the present context emphasizes. (b) Some other organisms are genuine artefacts or genuine partial artefacts – domesticated animals and plants are examples and so are genetically modified organisms. (See Lee 2005). (c) Zoo animals appear to form a distinct category of their own – these are not domesticated animals in the technical understanding of the term, but they are in all other ways subject to human manipulation and control as they are what one may call "immurated animals". To that extent, they are artefacts – see Lee 2006.

8. This is not to deny that there may be artefacts constructed by non-human beings, such as the dams of beavers. But such non-human artefacts are not germane to the line of argument pursued by this book. Nor is it to deny that when an artefact has outlived its original purpose, a new one may not be found for it. For instance, the few extant samurai swords are no longer used as weapons, as even the class of samurai no longer exists in Japanese society today, but they are, nevertheless, cherished as beautifully crafted objects as part of Japanese, indeed, even world patrimony.

9. For some details about the history of technology, the philosophy of technology and their relationship to the history and philosophy of science, see Lee 2008 or Lee 1999.

10. Lewis Mumford proposes a three-fold division in classifying the history of technology (whose edges are meant to be overlapping) in terms of the type of energy and characteristic materials used. The eotechnic phase is a water-wind-and-wood complex; the paleotechnic phase is a steam-coal-and-iron complex; the neotechnic phase is an electricity-and-alloy (as well as synthetic compounds) complex. The first, for him, stretches roughly from 1000 to 1750 CE, the second, from 1750 to 1850s, and the third, from 1850s to the present.

11. The classic definition insists on the criterion of moving parts; however, since the beginning of the electronic age from the last quarter of the twentieth century, this key criterion has now been dropped to include electronic devices, such as the computer as machines.

12. The escapement mechanism is said to constitute the beginning of the true mechanical clock. In China, during the Song dynasty, a clock using such a mechanism but powered by water rather than weights (which European clocks in later centuries relied on) had been devised by one called Su Song (1020–1101) in 1088. By the thirteenth century, engineers in the Islamic world had made numerous mechanical clocks.

13. On these points, see also Merchant (1980, 217): "... (watermills and windmills) were large, geared machines, that, as necessary parts of the new industries, were foci around which new forms of daily life became organized and institutionalized. With the spread of capitalist economic forms, these mills, together with furnaces, forges, bellows, cranes, and pumps, became an integral part of the everyday experience of many Europeans...." In other words, the ontological *volte-face* must be understood as part of the wider economic and social changes taking place in Europe towards the end of the Middles Ages to the seventeenth century.

14. In 1218, Genghis Khan broke off his attack on China and turned his attention to Central Asia and Europe. His grandson, Kublai Khan finally mounted the Chinese throne in 1279, establishing the Yuan dynasty which lasted till 1368. Marco Polo is said to have journeyed to China which he called Cathay. The Mongol rulers in China imported foreigners, such as Europeans and even some missionaries into the country and to its court in order to lessen its reliance on and the influence of the Chinese in its administrative machinery. Although the Mongol empire did not last very long, nevertheless, its impact on East-West exchange of ideas, technologies and artefacts was profound.

15. "No one man created the mechanical philosophy. Throughout the scientific circles of western Europe during the first half of the seventeenth century we

can observe what appears to be a spontaneous movement toward a mechanical conception of nature in reaction against Renaissance Naturalism" (1977, 30–31).

16. For a detailed study of Hobbes's contribution, see Lee 1989.

17. Ironically, Newton (1643–1727), acknowledged as the giant of the new (mechanistic) physics, even the greatest scientist of all times, engaged in no philosophy but indulged liberally in alchemy instead. Upon his death, when it was found that his alchemical activities and writings even exceeded those on physics, an attempt was immediately made to suppress such embarrassing evidence. So successful was the cover-up that the truth remained hidden for over two centuries until, by chance, the famous Cambridge economist, John Maynard Keynes, bought those papers in 1936 and revealed to the world their astonishing contents. Keynes ("Newton the Man" 1946) wrote: "... Newton was not the first of the age of reason. He was the last of the magicians, the last of the Babylonians and Sumerians, the last great mind which looked out on the visible and intellectual world with the same eyes as those who began to build our intellectual inheritance rather less than 10,000 years ago" at http://www-history.mcs.st-and.ac.uk/ Extras/ Keynes_ Newton.html .

    Robert Boyle (1627–1691), the famous scientist who was an atomist and whose name we come to associate with Boyle's Law, did not hesitate to accept that the body-is-machine; however, as he was a deeply religious person, he found the conflict between atomistic materialism and Platonic philosophy difficult to handle; he never satisfactorily resolved it – see King 1978.

18. Such a view may sound extreme, and although it is true that not all contemporaries or near contemporaries of Descartes swallowed it whole, nevertheless, it enjoyed widespread support till quite recently, as advocates of animal welfare and rights remind us. La Mettrie (1709–1751), for instance, while championing the view that man-is-machine, nevertheless, did not subscribe to the extreme Cartesian thesis that animals are plain automata.

19. http://www. cscs.umich.edu/~crshalizi/ LaMettrie/ Machine/.

20. As a matter of fact, by late Medieval times, even the imagination of theologians (apart from that of other elites as well as the aristocracy) had been captured by clocks and other mechanical automata. Certain authors had already referred to the cosmos as *machina mundi*. One of the most famous is Nicole Oresme (1323–1382), mathematician and theologian (Bishop of Lisieux). In 1370, he had written: "And these powers are so moderated, tempered, and ordered against their resistances that the movements are made without violence. And except for the lack of violence it is like the situation when a man has made a clock and lets it go and be moved by itself. Thus it was that God let the heavens be moved continually according to the proportions that the moving powers have to their resistances and according to the established order" (Merchant 1980, 223).

21. http://www.ucmp.berkeley.edu/history/paley.html.

22. We shall be returning to this thesis in Chapter 5.

23. As it turns out, the history of modern science and the history of modern technology do not exactly coincide chronologically. The latter, as we have said, started in the seventeenth century whereas the promised spin-off in

terms of technology did not occur till the 1840s. However, the ideological agenda was laid down at the very beginning of modern science itself. For details, see Lee 2005, Chapter 2.

24. So did Hume. He (*An Enquiry Concerning Human Understanding,* VII, II) wrote: "The only immediate utility of all sciences is to teach us how to control and regulate future events by their causes. Our thoughts and enquires are, therefore, every moment, employed about this relation."

25. The criticism may be advanced that such a possibility does not qualify as control at all in any sense of the term; if this criticism is conceded to be valid, then only the possibility of control in the strong sense remains as the goal of science. Indeed, the weak form of control may not be the truly desirable goal. Perhaps, it is *faute de mieux,* and at best, a prelude to the aspiration of controlling nature in the strong form. Being able to predict the onset of drought or rain is clearly better than not being able to do so at all. But it would be better if scientific theoretical understanding of meteorological phenomena ultimately enables one either to generate rain (when drought is undesired) or to hold rain at bay (when dry weather is desired). However, scientists do not give up so readily on the weak sense of control – meteorologists and vulcanologists work very hard to improve their respective models of prediction in the hope that more accurate prediction of good/bad weather, of volcanic eruptions would be welcomed as signs of progress. After all, lives as well as economic assets could be saved if improvements in prediction could be made.

26. On the points to be discussed, see also Kennington 1978.

27. http://faculty.uccb.ns.ca/philosophy/kbryson/rulesfor.htm.

4   Machines and Reductionism

1. This is not to deny that occasionally, though rarely, some could have emerged by happenstance, rather than by design – such as the telescope. Its invention is usually attributed to a Dutch spectacle maker, Johann Lippershey, c. 1605. Whether he was truly the first inventor remains uncertain; however, the key to the invention was the discovery that far away things appeared clear and nearby when two lenses, one convex, the other concave were accidentally placed behind each other. Galileo perfected the device in 1609.

2. "Manufacture" is used in the following senses: (a) in the original sense of being hand-crafted (to make by hand); (b) in the larger sense of any object made by humans using technology of any kind, whether traditional/simple or scientific/sophisticated.

3. This, today, occurs not only in the case of machines such as motor cars or fighter planes but also of computer software. The motives for doing reverse engineering may be diverse: when manuals have been lost; for military purposes; for commercial reasons, including finding out whether patent rights have been violated, and so on.

4. For instance, a building of historical importance standing, say, on a site which is now required for some other urgent use can, in principle, be dismantled stone by stone, beam by beam and then reassembled in exactly the way it was earlier structured on another site. Admittedly, a structure such

as a building is not a machine, but it has parts; any thing which has parts can in principle be dismantled component by component and later put back together again.

5. No serious exponent of holism today would hold the most extreme historic form of it (usually attributed to Hegel), namely, that wholes can exist totally independently and separately from the existence of their parts – that, should their components be all destroyed, the wholes would still exist in some Platonic heaven. The more defensible forms of holism would assert that, while wholes cannot exist in the absence of the existence of component parts, nevertheless, the relationships between wholes and parts at the level of methodology, metaphysics, logic, language, and so on cannot be adequately accounted for in the terms advocated by reductionist thinking. Furthermore, holists maintain that wholes can influence how the parts function, that downward causation (whole determining how parts perform), not merely upward causation (parts determining how whole performs) exists.

6. In the social sciences, the most famous recent pronouncement on this subject comes from two politicians in the 1970s and 1980s; namely, the US president, Ronald Reagan and the British prime minister, Margaret Thatcher. Thatcher has memorably said that there is no such thing as society, that there are only individuals. Ryle was eager to combat the legacy of Cartesian dualism which, as we have already seen, conceives the human being in terms of two substances – the body, which is a material, while mind is an immaterial substance. Ryle challenges the Cartesian view about mind, which in his opinion constitutes a category mistake, as mind is not substance at all. In contrast, the two leading politicians of the Western world in the last quarter of the last century might be said to have implicitly relied on reductionism to initiate their radical political programmes of transforming society from one which was society-led (in Britain and Europe rather than the United States) to one which endorsed individual citizens looking after their own interests. In other words, society-led intervention is woefully mistaken – society as an entity is not real and does not exist. To hold that society exists and can have goals and purposes over and above those held by individuals is to commit a category error.

   Hobbes, in the seventeenth century, had already advocated (in thought) the dissolution of society into its components – that is, individuals using the method of what is called in this book the principle of "reverse engineering." He wrote (*Leviathan*, Chapter 4):

   > For everything is best understood by its constitutive causes. For as in a watch, or some such small engine, the matter, figure, and motion of the wheels cannot well be known except it be taken asunder and viewed in parts; so to make a curious search into the rights of states and duties of subjects, it is necessary, I say, not to take them asunder, but yet that they be so considered as if they were dissolved.

7. The thesis of emergence for the sake of brevity is articulated here in its most simplistic form. Later chapters will explore some aspects of it in greater detail.

8. Aspects of this claim will be looked at in greater critical detail in the rest of the book.

5   Organism A Machine

1. For a systematic and detailed exploration, see Lee 2005, in particular, Chapter 3.
2. One must note that there are two meanings of "hybrid" and "hybridization". In its original form, as used here, it simply means the "cross breeding" or sexual combination of two varieties of plant or animal. A hybrid is no more than the product of such a union. However, after 1930, when Mendelian genetics had spawned its own technology, "hybridization" acquired a much narrower meaning, referring to the outcome of combining two in-bred lines, as in "hybrid corn".
3. See Mumford 1967, 151.
4. However, this does not necessarily mean that the end product, a new variety or an improved variety, is the result of deliberate selection on the part of the farmer/breeder. There are other distinct possibilities involved – see Rindos 1984, 1–9.
5. But propagation by cloning is, indeed, also an ancient practice. A few fig clones survive today from the classical Roman period – see Simmonds 1979, 127.
6. According to Jared Diamond (1999, 57–75), the selection of desirable characteristics of animals involves six criteria: flexible diet; reasonably fast growth rate; tendency to breed under captivity; generally non-aggressive; a cool head, not readily panicked; willingness to accept humans as head of their social hierarchy.
7. For simplicity's sake, the past tense is used with regard to pre-Mendelian agriculture. However, it remains true there are parts of the world (admittedly shrinking) which still practise peasant, craft-technology agriculture. Furthermore, one should bear in mind that the land-races still in existence in developing countries today are biotic artefacts (but with a much lower degree of artefacticity) just as the hybrids of Mendelian technology and the genetically engineered plants (and animals) of biotechnology are biotic artefacts. The land-races are not raw germ-plasm.
8. However, such an assessment may be somewhat flippant and appears to have overlooked the superbly sophisticated results or effects of such a method of domestication – just consider the maize (corn) as opposed to its wild ancestor, teosinthe, as well as the numerous varieties of dogs, all bred from the same wild ancestor, the wolf.
9. Unfortunately, Darwin did not fancy reading some obscure publication written, to boot, in German; he did not bother even to cut open the pages of the article. As a result, Darwin relied for his genetics in his theory of natural selection on Lamarkianism (which held that characteristics acquired in an individual's life time, such as housemaid's knee, can be transmitted from parent to offspring). However, later, Neo-Darwinism ditched Lamarkianism, incorporating Mendelism instead.
10. Other plants, if chosen as experimental subjects, would have presented genetic complexities which would have muddied the waters and prevented him from formulating such neat and tidy laws.
11. This term is coined after the re-discovery of Mendel's work to refer to the genetic endowment of an individual organism, while the term "phenotype" refers to the organism's manifest characteristics.

12. See Olby 1966, 77–8.
13. For details about the traditional practice of "single line selection" and the new practice of double cross hybridization, see Lee 2005, 155–6.
14. For details, see Lee 2005, Chapter 4.
15. Two comments are called for here:
    1. It is not intended to deny, that what exactly counts as a species, is uncontroversial in biological literature today. However, for the purpose of biotechnology, what is crucial is that, in absolutely clear cases, where the organisms are recognized to belong to two distinct species, such as humans and mice, it is possible to insert genetic material belonging to the former into the latter, and *vice* versa – a phenomenon, which does not occur naturally, but only as a consequence of this particular form of advanced technological intervention.
    2. This is a matter of terminology. In this book, "transgenic organism" is used to refer to an organism into which genetic material from another organism belonging to a totally different species/kingdom has been inserted. In the opinion of this author, another term "chimera" should be used to refer to other very different kinds of genetically engineered entities, such as those which combine cells from different fertilized eggs but all belong to the same species.
16. For a more detailed discussion on this point, see Lee 2005, Chapter 6, section entitled, 'The Humanisation of Biotic Nature: The Supersession of Natural Evolution'. Another historic step has been taken in May 2010 when Craig Venter and his scientists claimed that they had succeeded in creating a synthetic genome – see http://www.sciencemag.org/cgi/content/full/328/5981/958; http://www. guardian.co.uk/science/2010/may/20/craig-venter-synthetic-life-genome
17. See Maturana et al. 1974; Varela 1979; Maturana and Varela 1980, and for a non-technical presentation of this work, see Maturana and Varela 1988, especially Chapter 2.
18. The concept may be traced, for instance, to systems analysis, since 1980; it is also related to the notion of dissipative structures as introduced by the physicist Prigogine and others in 1984.
19. One would like to emphasize the force of "implicit" in this context. Whether Maturana and Varela have done this consciously or not is irrelevant; what is significant is that their account reflects so admirably the scientific-cum-technological goal of molecular genetics and molecular biology.
20. While genetic engineers would welcome such an account, it is not obvious that biologists interested in natural evolution would, as to make sense of the subject matter of their study, they cannot dispense with the thesis of intrinsic/immanent teleology.
21. One historian of medicine, Hans-Jörg Rheinberger (2000, 25) has written:
    The molecular biologist, as the molecular engineer, today, has long abandoned the working paradigm of the classical biophysicist, biochemist or geneticist. He no longer constructs test tube conditions under which the molecules and reactions occurring in the organism are analyzed. Instead he constructs objects, that is, basically, instruction-carrying molecules which no longer need to pre-exist within the organism. In reproducing them, expressing them, and screening their effects, he uses

the milieu of the cell as their proper technical embedding. The intact organism itself is turned into a laboratory. It is no longer the extracellular representation of intracellular process, ie., the 'understanding' of life that matters, but rather the intracellular representation of an extracellular project, ie., the deliberate 'rewriting' of life. From an epistemic perspective, this procedure makes the practice of molecular biology, qua molecular engineering, substantially different from traditional intervention in the life sciences and in medicine. This intervention aims at *reprogramming* metabolic actions, not just interfering with them.

6 Human Organism is Machine: MEDICINE

1. Today when funding for scientific research (in the West) is increasingly scarce, scientists have become less coy in expressing the wish that their "bluest" of "blue sky" research may also prove a technological winner. No doubt, the backers of the Large Hadron Collider (the LHC or the Big Bang Machine) underneath Geneva fervently hope that finding "God's particle" would lead to spin-offs, which could power the next spurt in economic growth, the next Kondratieff wave. Cold fusion as a possibility refuses to lie down as the stakes are too high. Should a version become available which is free of theoretical incoherence as well as susceptible to practical application, then a new source of energy would become available and the pioneers would be awarded the Nobel Prize several times over.
2. See Lee 1999.
3. For a more detailed discussion of this perspective, see Lee 2008.
4. Descartes says: *cogito, ergo sum* – I think, therefore I am.
5. Biotechnology is also called genetic engineering; scientists working in the field proudly call themselves genetic engineers. Chapter 8 (in its first section) will explore this in greater detail. See also section in Chapter 7 on genetic therapies in medicine.
6. The subjects respectively listed under Pre-clinical and Clinical training reflect in general the pattern of medical education in the UK, rather than, perhaps, in other parts of the world.
7. The term also refers to other areas of medical investigation which are not germane to the pre-occupation of this book.
8. Today, computer simulation is an additional, if not, exclusively alternative tool.
9. Sometimes the researchers may try to surmount this constraint by volunteering themselves. As will be discussed in a later chapter, modern medical trials methodologically require a large number of patients – a volunteer of one or a few is not to the point. That is why in some cases animals (those as near to humans as possible) are used instead, although this strategy, too, has come under moral (as well as methodological) criticism. It also explains certain episodes in the history of World War II where the results of Nazi or Fascist science were gratefully accepted by the medical/political establishments of the victorious Allies. Take the notorious case of the experiments (on biological warfare, in general) conducted by Japanese scientists of Unit 731 on primarily (though not exclusively) Chinese civilians and prisoners-of-war which involved, in particular, vivisection – no anaesthetic was given

as this would have interfered with the results. A notorious example of such a type of scientific investigation studied the impact of cold on human physiology. Water was poured over the outstretched limbs of the victims which would freeze over in the extreme cold of north China in the winter. The investigators would tap the limbs, telling by the sound whether they had frozen thoroughly. (The victims were suitably referred to as "logs"; but also sometimes as "monkeys".) Hot water was next poured over the "logs" to thaw them out before dissection. The United States, upon discovering this and similar activities, hushed up the affair in exchange for the invaluable results obtained from the Japanese scientific investigations. These results are indeed considered invaluable for two reasons: (i) the Americans realized that they themselves could not undertake such a type of study without incurring moral opprobrium; (ii) they helped crucially to advance physiological understanding of hypothermia in the human organism such that many of today's effective techniques for saving people thus exposed were developed in the years after World War II. (See http://www.copi.com/articles/guyatt/unit _731.html; Harris 1995.) See also later discussion about the ethical constraints on conducting RCTs in epidemiology in Chapter 12.

10. In the UK, recently, a body has been set out called the National Institute for Health and Clinical Excellence (NICE) a key remit of which is to determine this matter.

11. Numerous senses of the term "natural" or "nature" must be distinguished. The sense used in this context is the opposite in meaning of "supernatural". The sense of "natural" referred to earlier in the book is the ontological foil of "artefactual". For details on the various senses of "natural", see Lee 1999.

12. A term has even been coined to refer to such a kind of human biotic/abiotic hybrid – it is "cyborg". It first appeared in the 1960s in science fiction discourse. However, it is also used to denote real cyborgs, that is to say, patients whose lives are being prolonged or whose quality of life is enhanced through the result of sophisticated surgical/technological interventions.

13. Surgery will be examined in detail in Chapter 8.

14. For details, see Lee 2005, Chapter 2.

15. See Lee 2005, Chapter 2.

16. Other bed-side diagnostic methods included inspection, palpation and percussion.

17. One historian of medicine (Watts 1996, 360–1) has aptly put these points as follows:

Along with orthodox medicine goes a machine model of the human body. Where kidney disease might once have been considered the consequence of evil spirits, wicked deeds, a malicious deity or some other such influence, it is now viewed as a material problem: a failure of the biological that should be filtering, cleaning, and adjusting the chemical make up of the body's fluids. The renal physician is neither a priest nor a sham but the physiological equivalent of a domestic plumber. And so it is with most other branches of medicine from gastroenterology to gynaecology; doctors are trained, primarily, as technicians skilled at diagnosing and fixing failed body mechanisms. To pursue this demanding trade they need sophisticated equipment such as brain scanners, fetal monitors, endoscopes, lasers, radioactive chemicals, artificial hearts, and computers. "... The doctor...is a body technician. ..."

7   Biomedicine: Some Sciences

1. Hahnemann had intended it to be a term with unflattering overtones, as naturally, he considered homeopathy to be a superior type of medicine altogether. However, since its introduction, the term is now used even by the orthodox medical establishment in certain contexts in a purely descriptive way without its original negative overtones. Allopathic medicine is defined as: "The system of medical practice which treats disease by the use of remedies which produce effects different from those produced by the disease under treatment. MDs practice allopathic medicine." In contrast, homeopathic medicine and its therapy are based on the concept "that disease can be treated with drugs (in minute doses) thought capable of producing the same symptoms in healthy people as the disease itself" – see "Definition of Allopathic medicine" at http://www.medterms.com/script/main/art.asp? Article key = 33612.

2. Some lists also mention Japanese Kanpo, but this, in the main, appears to be heavily dependent on Chinese medicine.

3. For a full discussion, see Gaines and Davis-Floyd, 2003 at http://www.davis-floyd.com/ userfiles/Biomedicine.pdf.

4. For details of the relation between craft-based technology and science on the one hand and science and technology on the other, as well as between philosophy of science and philosophy of technology, see Lee 2005, Chapter 2; for only a brief discussion about the relation between deeper scientific theories and deeper technology, see Chapter 8 (of this book).

5. For those interested in the humoral theory of disease, the literature is rich. Suffice it here to mention only a few. For a very short account, see http://ocp.hul.harvard.edu/contagion/humoraltheory.html; for a longer account see Wootton 2006; for a sympathetic account, see Osborne 2009 at http://www.Greekmedicine.net/Principles_of_Treatment/Introduction_to_Therapeutics_in_Greek_Medicine.html .

6. However, cautery (which involved applying a hot iron to parts of the body) more or less dropped out of the repertoire earlier, by the Renaissance; but this did not prevent René Laennec (the famous physician who in 1816 invented the stethoscope) to use it on patients suffering from phthisis (called tuberculosis today) – he used a hot copper rod to burn the chest in 12–15 places.

7. A case of the famous who was bled to death by his physicians in 1799 was George Washington – see Moerman 2002; http://en.wikipedia.org/wiki/Bloodletting. Another was Charles II who died of a relentless treatment based on bleeding and emetics to which latter mixture was even added 40 drops of an extract of human skull, and so on – for an account, see Evans 2004, 130–1.

8. A scratch was first made on the skin, then a cup was applied to it – air was either pumped out from a hole in the bottom of the cup or by first heating the cup before putting it over the skin. This method was regarded as a gentler method than venesection in drawing blood from a patient.

9. The French favoured leeches so much that during the nineteenth century, they had run out of them and had to import (from Turkey), increasing from a mere 100,000 in 1824 to 33 million in 1827; England in turn imported

six million from France. The demand was just as brisk in the rest of Europe. (See Duke 1991.) Today, there is renewed interest in leeches, but their use is severely restricted to certain very specific contexts such as in reconstructive and plastic surgery; research is also looking into the anticoagulant produced in the saliva of some leeches.

10. As the humoral theory held that the four humours were found in blood, to get a proper balance between them in the body, it follows that any excess could and should be eliminated via blood letting. When venesection was employed, the doctor used a lancet to open up a vein – for this reason, one of the world's leading medical journals is called *The Lancet*. Indeed, as late as 1911 *The Lancet* carried an article entitled "Cases illustrating the uses of venesection", which included high blood pressure and cerebral haemorrhage. Indeed, blood-letting through two millennia had been invoked to treat nearly every disease, ranging from acne, asthma, diabetes, fever, gout, poisoning (including carbon monoxide and mustard gas suffered by the victims of such gas attacks in the trenches of the First World War in 1916) to being a general regimen for maintaining health and longevity. (For more details on these points, see Wootton 2006; Seigworth 1980; http://www.pbs.org/wnet/redgold/ basics/ bloodletting history.html; http://www.pbs.org/wnet/redgold/basics/bloodlettinghistory2. html; http:// www.pbs.org/ wnet/redgold/ basics/bloodlettinghistory3.html.

11. A laboratory is here used in this larger sense of the term – it is any space designed for the purpose of a specific scientific investigation to be conducted under the strict conditions laid down by the new methodology.

12. We shall see later in Chapter 12 that not all medical research and investigation can meet such a demanding methodological norm. As a result, there is at least one whole area which appears to suffer, for want of a better word, what may be called "a Cinderella status", for which so far no Nobel Prize Committee in Medicine and Physiology has bestowed an award.

13. On the reductionist character of anatomy, see, for instance, Amsterdamska and Hiddinga 2003.

14. Thomas Sydenham (1624–89), considered to be the father of modern English medicine and sometimes called the English Hippocrates, was firmly of the opinion that anatomy as a science could have no relevance to medicine – see King 1978. To put Sydenham's point with greater caution and less provocatively, one could say that anatomical studies had nothing to offer medicine as "succour". For Sydenham, the foundation of medicine was clinical or bedside medicine, based on accurate/objective observations of the patient's symptoms, and not on anatomy or physiology as medical sciences.

As the science of anatomy developed and evolved, anatomists went beyond merely the study of parts of the skeleton and the locations of the various internal organs. Anatomy turned into pathology. The anatomist-cum-pathologist then looked out for lesions which could be correlated with the disease/illness for which the physician had diagnosed the patient when the latter was still alive and under his care. (This advance occurred with the publication of Morgagni's *Seats and Causes of Disease* 1761, considered to be a milestone both in the history of pathology and in the theory of explanation – see King 1978, 194.) This improved greatly medical understanding but in itself, pathology, too, did not have anything really to offer medicine as succour.

15. The first anatomical theatre was built at the University of Padua in 1594; Leiden had its completed in 1596, and Bologna's was designed as such in 1637. (http://www.ctsnet.org/doc/9609 shows a picture of the Padua theatre.)

16. Descartes read Harvey's *De Mortu Cordi* in 1632 and accepted his account of the circulation of the blood (though not other findings). Blood circulation fitted in nicely with the Cartesian view of motion with the single motion of the heart as the source of all the other motions of the body. He wanted to convince the theologians who adhered to the old philosophy that the new science and the new philosophy were perfectly compatible ultimately with the existence of God. He wanted to show that the new science and the new philosophy would lead naturally to a new medicine; it was not the case that only Aristotelianism would naturally lead to medicine. Above all, he meant his *Treatise on Man* to create an impact more at the philosophical rather than at the medical level. French (1989, 52) writes: "... Descartes had undertaken a campaign to depose Aristotle as The Philosopher and to occupy his place, not only for the reading public, but in the universities."

17. For a detailed account of the differences between observation and experiment, see Bernard 1957, part one, chapter 1.

18. "On n'a rien écrit de plus lumineux, de plus complet, de plus profond sur les vrais principes de l'art si difficile de l'expérimentation", *Moniteur Universel*, 311 (7th novembre 1866) 1284-5; http://www.bookrags. com/biography/ claude-bernard/.

19. These included: the functions of pancreatic secretion, the glycogenic function of the liver, vasomotor functions, the trophic effects of nerves, the effects of drugs such as curare, through his concept of *milieu interior* (the internal environment) which led eventually to our understanding of homeostasis, the discovery that red blood cells contained a chemical which carried oxygen. (This chemical was later identified by E. F. Hoppe-Seyler in 1857 as haemoglobin.)

20. See Bernard Cohen's Forward to the Dover edition of Bernard's classic in 1957, nearly a century after its original publication.

21. "Theories are only hypotheses, verified by more or less numerous facts. Those verified by the most facts are the best, but even then they are never final, never to be absolutely believed" (Bernard 1957, 165).

22. See Popper, 1959 and 1963. For a brief account of Popper's writings, see Thornton 2009 at http:// plato.stanford.edu/entries/popper/. For a critical account of Popper's philosophy of science, see Lee 1985, 175–83.

23. Bernard 1927 and 1949, 56.

24. Alas, this clear-sighted methodological rule is violated time and time again even today, especially in pharmacological research, financed in general by the pharmaceutical industry.

25. Bernard 1927 and 1949, 38.

26. Bernard himself appeared to be aware of the point made here. He wrote (1976, 78): "Le cadavre est l'organisme privé du mouvement vital, et c'est naturellement dans l'étude des organes morts qu'on a cherché la première explication des phénomènes de la vie, de même que c'est dans l'étude des organes d'une machine en repos qu'on cherche l'explication du jeu de la machine en mouvement." This author's rough translation reads: "The

corpse is an organism without motion, and it is naturally in the study of dead organs that one finds the first explanation of life phenomena, in the same way that it is in the study of the 'organs' of a machine which is at rest that one finds an explanation of the 'spirit' of the machine when it is in motion."

Bernard unhesitatingly accepted the ontological *volte face* that the human organism is living machine – see Bernard 1957, 76–80. However, Bernard did not subscribe to crude reductionism but had a more nuanced account of the relationship between parts and whole. He wrote (1957, 89): "We really must learn...that if we break up a living organism by isolating its different parts, it is only for the sake of ease in experimental analysis, and by no means in order to conceive them separately. Indeed when we wish to ascribe to a physiological quality its value and true significance, we must always refer it to this whole, and draw our final conclusion only in reaction to its effects in the whole." He went on to say that it was crude reductionism which gave vitalism a convenient stick to beat those including himself who were resolved to undermine such an approach.

27. For details of the chemical reactions involved see http://www.chemistryexplained.com/Va-Z/W-hler-Friedrich.html;http://en.wikipedia.org/wiki/W%C3%B6hler_ synthesis.
28. For a short but succinct account, see Hoefer 2008.
29. In particular, see Bernard 1957, 138.
30. See Temkin 2002, Chapter 8.
31. For a brief account, see "Gene Therapy" at http://www.ornl.gov/sci/techresources/ Human_Genome/ medicine/genetherapy.shtml.
32. Haemophilia is a sex-linked condition leading to uncontrollable bleeding owing to the absence of one of two protein co-factors necessary for the coagulation mechanism to function properly. There are two forms of haemophilia A and B. Haemophilia B is due to factor IX deficiency, while Haemophilia A is due to factor VIII deficiency and is nine times commoner than Haemophilia B. As far as common usage is concerned, the "haemophilia" is confined to the A variety and the B variety is called "Christmas Disease" (after the first patient identified as the bearer of the form).
33. Doctors from the *Universitat Autonomia de Barcelona* and researchers from the Cefer Institute of Reproduction in Spain have recently published in the journal, *Prenatal Diagnosis,* about the case of a Spanish woman who is a haemophilia carrier, but who has chosen not to have daughters. The medical team involved made sure that the embryos implanted in the woman's uterus were male. (The technique used is PGD, pre-implantation genetic diagnosis.) This goes beyond the stage sanctioned by the British HFEA (Human Fertilisation and Embryological Authority) which allows the elimination of embryos possessing the defective gene for haemophilia before transplantation only in the case where the mother is a carrier; it does not yet permit a woman or couple to choose the sex of an embryo to ensure that no daughter would be born, who might eventually pass the defective gene to her sons, when she in turn reproduces. (See Meek 17 October 2000, 6). A case of success has occurred with regard to cystic fibrosis amongst Jewish Ashkenazi people through genetic counselling when both the male and female partners are carriers of the disease. Should they have a child, there is a 25 per cent chance that the

offspring would have cystic fibrosis, a 50 per cent chance that it would be a carrier like themselves, and a 25 per cent chance that it would neither be affected nor a carrier – this is in accordance with Mendel's laws of inheritance. Parents, when presented with the results of testing that the offspring would be a sufferer, would normally decide to have an abortion. See http://www.jewish virtuallibrary.org/ jsource/ Health/ genetics.html.

However, one must bear in mind that spontaneous mutation in Haemophilia is quite high and cases occur in people with no genetic history of the condition are quite common – hence using the method described above cannot eliminate Haemophilia for all times.

34. The same situation obtains with regard to the sex of embryos – some countries permit the use of pre-natal diagnosis of the sex and then to abort the foetus if the sex is of the "wrong" kind (usually female, though not invariably so).

35. See Collins, 2010.

36. For a brief account, see http://www.ncbi.nlm.nih.gov/About/primer/snps. html .

37. See http://las.perkinelmer.com/content/snps/genotyping.asp#human ; Marchant 2000, 46–50.

38. Of course, this new science of genomics is not merely relevant to the genetic factors involved in causing disease in the human population, but also to improving yield in food crops and in animal husbandry. The economic horizon is, indeed, a very wide one.

39. The tests for the drugs involved and their dosage were not in the main designed with them in mind – see Healy 2009, Chapter 8 for some obvious pitfalls in such a situation. There are, however, useful guidelines to help prescribers adjust dosage for child patients, such as the well-known Harriet Lane Handbook produced by the Johns Hopkins University – see http://www.mdconsult.com/das/ book/body/178794899-2/0/1755/0.html. Furthermore, the recruitment of paid volunteers for drug trials in general are addressed to the age group 18–85 years – see, for instance, http://www. gpgp.net/. Under pressure, the United States government passed a federal law in 1997 permitting pharmaceutical companies very generous terms for exclusive marketing of their drugs in exchange for conducting drug trials on children – see, for instance, http://www.ahrp.org/infomail/1002/18.php ; http://www.gpgp.net/ ; for cancer treatments in the UK, see http://www. nature.com/ bjc/journal/ v88/n11/full/6600990a.html ; for paediatric clinical trials and their related difficulties, see, for instance, http://linkinghub. elsevier.com/ retrieve/pii/S0022534701684246.

40. The ethical constraint could be removed today in the case of animal experimentation via cloning, as cloning produces the equivalent of monozygotic twins. However, society appears not to find it morally acceptable to permit (a) tinkering with the individual human genome, (b) cloning such tailor-made/modified individuals for the purpose of fulfilling the demands of RCT. Chapter 12 of this book looks in further detail at the ethical constraints imposed on RCTs in epidemiology.

41. Another candidate may be the phenomenon of the placebo effect, which will be explored in Chapter 8. However, the present book cannot develop this theme in detail.

8   Biomedicine: Some Technologies

1. The same holds true of pre-biomedical therapies. For instance, for cupping, one would use a cup, for cautery, one would use a metal rod, for venesection, one would use a lancet, and so on.
2. He coined the word "cell".
3. For details, see Mazzarello 1999.
4. For the purpose of this discussion, the distinction between technique and technology may be made as follows: technique basically is free standing without technological input, although many techniques also involve technology. The technique of singing *per se* involves only using one's lungs, the diaphragm, larynx, mouth, other relevant parts of the body in certain ways in order to project certain sounds. However, today, even opera singers in some of the world's leading opera houses no longer simply rely on techniques of singing alone – some may even carry hidden microphones on their bodies. Technologies necessarily involve instruments and artefacts, but techniques *per se* do not.
5. Furthermore, patients who undergo frequent blood transfusions tend to accumulate an excessive amount of iron, which causes damage to the heart and liver, as well as often interfering with normal growth and development.
6. The latest development on this front, reported in late September 1997, is the success of the laboratories of the American Red Cross in Rockville in producing factor VIII in pig's milk. The pigs have been genetically modified to do so. The scientists have injected pig embryos with an artificial version of the human gene responsible for the liver in making factor VIII. To ensure that the blood-clotting protein would be found only in the pig's milk and nowhere else, the human gene has been tied in with a pig gene, which only works in its mammary glands. (See Coghlan 1997, 10.)
7. Another example of the same progression at work concerns the condition called Gaucher's disease. Philippe Gaucher discovered it in 1882. It is inherited from two carrier parents who themselves may be free of the symptoms. The sufferer's body is unable to break down the chemical glucocerebroside, found in the membranes of white and red blood cells, which enables macrophages containing fatty glucocerebroside globules to accumulate in the liver, blood marrow and spleen. This could lead to brittle bones as well as the liver and spleen swelling up. In the 1980s, Dr Roscoe Brady of the National Institute of Health (USA) identified the enzyme, which the patient lacks, responsible for such symptoms. He managed to extract the critical enzyme from placentas, and administer it as a drug to patients. This first-generation drug, called ceredase, is manufactured by a company called Genzyme. But it can only be expensively produced. Dr Brady, in the late 1980s, went on to identify the gene that makes the enzyme, which breaks down glucocerebroside. This gene is then inserted into cells isolated from Chinese hamsters. The cells are grown in vats producing unlimited amounts of cerezyme, the biotechnological version of ceredase. In 1994, Genzyme was on the verge of marketing this second-generation drug with the expected approval of the US Food and Drug Administration. Already the next new-generation product on the horizon is being put in place, a device to insert the actual missing gene into the patients' bodies. (See McKie 1994.)

8. This rule can now be instantiated by means of Pre-implantation Genetic Diagnosis (PGD), using *in vitro fertilization* to grow embryos outside the uterus, then testing to check that they do not carry the gene for the disease, and implanting only such embryos. In October 2000, an American couple from Colorado – Adam and Lisa Nash – announced that their infant son, born in August of that year, had been thus conceived – with the precise aim of harvesting cells from his umbilical cord for infusion into his elder sister, who was suffering from a rare inherited genetic disorder called Fanconi anaemia, a condition which stops cell production in the bone marrow. The medical team, just after mid-October, declared the procedure had been successful and that the Nash daughter had been saved. Shortly following this report, University College Hospital, London announced that it, too, would be using PGD to ensure that a family afflicted for generations with a form of bowel cancer known as familial polyposis (which killed half of those who inherited the condition in their early middle age) would escape the condition. (http://www.dh.gov.uk/en/Publicationsandstatistics/Publications/PublicationsPolicyAndGuidance/DH_4118934, 2002).
9. For more detailed arguments, see Lee 2006, Chapter 2.
10. Neolithic peoples throughout the world performed craniotomy, drilling holes into the top of the skull, probably to treat head injuries, epileptic fits, migraines, and so on – see Sperati, G. 2007, "Craniotomy through the ages." http://www.ncbi.nlm.nih.gov/pmc/articles/PMC2640049/; Ellis, Harold 2001, *A History of Surgery*. (London: Greenwich Medical Media, 2001). (However, some scholars think that the operation was for a religious purpose, to let out some evil spirit residing in the patient.)
11. In the history of medical therapy, one could even say that surgery since the seventeenth century had more to offer than internal medication through drugs. As we have noted in an earlier chapter, the latter, based on the treatments of humoral medicine, did more harm than good before the advent of modern drugs and techniques from the 1840s. This reality led to a period of what is sometimes called "therapeutic nihilism" – see Temkin 1997.
12. In London, by 1308, an organization, which later became known as the Barbers Company, was set up to regulate the affairs of its members. These included surgeons. In 1368, the Guild of Surgeons was formed, followed eighty years later by the Guild of Barbers. Under an Act of Parliament in 1540, the two guilds were amalgamated; the amalgamation lasted roughly two centuries. See http://www.barberscompany.org/historical_group.html#The%20history%20of%20the%20company.
13. He not only excelled in treating battlefield wounds, but also in obstetrics. He served four Valois kings – see Lawrence 1993; http://wapedia.mobi/en/Ambroise_Par%C3%A9; http://www.sciencemuseum.org.uk/broughttolife/people/ambroisepare.aspx.
14. http://wapedia.mobi/en/History_of_surgery; Hughes Evans, "History of Surgery" at http://www. slideworld.com/slideshows.aspx/The-History-of-Surgery-ppt-560577.
15. http://members.rediff.com/bloodbank/History.htm.
16. Theoretical understanding of different human blood groups, however, did not occur till the 1900–01, although antisepsis already was in place by the

last quarter of the nineteenth century. See http://mahasbtc.aarogya.com/ index.php/history-of-blood-transfusion.

17. http://generalmedicine.suite101.com/article.cfm/the_history_of_anesthesia; http://www. Discoveriesinmedicine.com/Enz-Ho/Ether.html; http://priory.com/homol/History% 20of%20 Anaesthesia.pdf.

18. See Ellis 2001, Chapter 8.

19. The split in France also took place around the same time; a royal charter established the Académie de Chirugie in 1731. In 1743, Louis XV severed the link between the surgeons and barbers – see Porter 1996.

20. For a detailed account of the surgery of warfare, see Ellis 2001, Chapter 9.

21. Of course, there is a very sound practical reason why surgery should be so closely involved with anatomy, as without anatomical knowledge, a surgeon would wreck havoc on the patient on the operating table – see Temkin 1977.

22. http://www.pedisurg.com/PtEducENT/tonsils.htm. Compare this more extreme view with a more moderate one issued by the NIH - http://www. nlm.nih.gov/ medlineplus/ tonsilsand adenoids.html#cat1. However, in other countries, the attitude to tonsillectomy is somewhat different – doctors, on the whole, would not recommend the operation in advance, but only in the light of a personal history of frequent and severe attacks of tonsillitis. Furthermore, they also believe that tonsils are not without function as they believe that they help fight ear, throat and nose infection in young children – see http://hcd2.bupa.co.uk/ fact_sheets/html/ tonsillectomy. html. Theodor Kocher obtained the Nobel Prize in 1909 for his work on the physiology, pathology and surgery of the thyroid gland. Thanks to his study of the serious consequences of surgically removing the thyroid gland (total strumectomy), which helped to cast light on its normal functions, total strumectomy was discarded and many were saved from severe suffering as a result. Furthermore, by the 1890s the isolation of active thyroid hormones made replacement therapy possible. See http://nobelprize.org/nobel_prizes/ medicine/ laureates/1909/press.html.

In the same spirit, the adrenal gland was also surgically removed. But what is perhaps even more surprising (at least to readers today) is that at the turn of the twentieth century, Elie Metchnikoff (1845–1916), a well-known member of the Pasteur Institute, Nobel Prize winner in 1908 for his work on the immune system, propounded the view (1903, 249 & 252) that the large intestines in the human organism had contributed "nothing to the well-being of man", but only posed many dangers to it. As it would take too long for evolution to attrite that organ, he proposed it best to accelerate that effect by using surgery to remove most of it – see Albury 1993.

23. See Ellis 2001, Chapter 15.

24. For one account of the history, see Temkin 2002, Chapter 8.

25. Today, it is recognized that there are 16 main groups of active ingredients, of which the alkaloids is one – for details see http://www.health24.com/ natural/Herbs/17-666-674,22561.asp.

26. Ipecac is itself derived from the bark of the roots from the plant, ipecacuanha, a native of Brazil. Emetine is a powerful poison. See "Ipecac", *Columbia Encyclopedia* 2008 – http://www.encyclopedia.com/topic/ipecac. aspx.

27. Up to then, the bark was dried, then ground up into a fine powder which was mixed with a liquid, usually alcohol, and drunk. In general, the new biomedically produced drugs are in the form of pills.

28. Yet another claim is that the synthesized drug would qualify for a patent which would enable pharmaceutical companies to recuperate the large amount of money put into finding new drugs for diseases. In some cases, a very different kind of benefit is made – resorting to the synthetic chemical would prevent the plants from environmental threats including their extinction.

29. For the molecular structure of quinine, see http://www.pharmacy.wsu.edu/ History/ history24.html; for a brief account of the synthesising of other active ingredients such as aspirin (willow bark), taxol (yew), hypericin (St John Wort), see http://www.oum.ox.ac.uk/educate/resource/ miracle.pdf.

30. Wolff 1995, 4.

31. This was the 914th arsenical substance in the list of substances explored in his research programme. This drug, though less effective in curative terms than its nearest product such as salvarsan, is, nevertheless, more readily manufactured, more soluble and therefore more easily administered – see http://nobelprize.org/nobel_prizes/medicine/laureates/1908/ehrlich-bio.html.

32. See BBC2 2010, Chapter 2.

33. For a philosophical assessment of nanotechnology, see Lee 1999 and 1997.

34. See http://nihroadmap.nih.gov/nanomedicine/ and its associated websites; Jain 2008, Chapter 4.

35. For quick account of: oestrogen, see http://www.guardian.co.uk/lifeand-style/ besttreatments/ anorexia-oestrogen-its-special-role; for the role of progestins (the term for natural and man-made progesterone) in the combination birth control pill, see http://contraception. about.com/od/ thepill/p/Progestins.htm. See also http://www.medic8.com/ complementary/ oestrogen-progesterone.htm.

36. See BBC2 2010, Chapter 5; http://www.nhs.uk/conditions/Male-contraceptive-pill/Pages/ Introduction.aspx.

37. See http://www.kirjasto.sci.fi/parace.htm.

38. Denying him this honour is really analogous to denying Newton the title of "founder of modern physics" because of his equal obsession with alchemy.

39. See Healy 2002, 38–41; http://www.chemheritage.org/EducationalServices/ pharm/chemo/ readings/ehrlich.htm; http://www.chemheritage.org/ EducationalServices/pharm/chemo/ read ings/ehrlich/pabio.htm.

40. http://nobelprize.org/nobel_prizes/medicine/laureates/1988/press.html.

41. Schardein 1975 as cited in Hawkins 1983, 17–8.

42. Sharpe 1988, 107.

43. BBC2 2010, Chapter 6. In the UK, GlaxoSmithKline (GSK) alone, between 2005 and 2009, eight drugs had been withdrawn, and several had been found to have serious side effects, such as Seroxat (an SSRI); Seroxat has now been acknowledged to raise the risk of suicide eight times. See also Weatherall 1996.

44. On these points, see BBC2 2010, Chapter 10.

45. This law is often expressed as The Therapeutic Index (TI) – if 50 out of 100 rats given a drug for a disease die and 50 survive fitter than before, the TI is

1 and the experimental drug is held to be unacceptable for use with regard to human beings.

Evans (2004, 120) has given another version of the first law of pharmacology: "In fact, one might propose the principle that the more effective a drug is, the more likely it is to have powerful and potentially harmful side-effects, as the first law of pharmacology."

46. Lock 1997, 138; Evans, Thornton and Chalmers 2007, 52; http://www.james-lindlibrary. org/ pdf/testing-treatments.pdf.

47. BBC2 2010, Chapter 6.

48. http://www.britannica.com/EBchecked/topic/323108/Emil-Kraepelin; http://ajp.psychiatryon line.org/cgi/reprint/163/10/1710.pdf.

49. http://findarticles.com/p/articles/mi_g2699/ is_ 0005/ai_2699000523/; http:// www. bipolarworld.net/Bipolar%20Disorder/History/hist3.htm; http://www.economicexpert.com/ a/ Emil:Kraepelin.html. For a general account of the history of psychopharmacology, see René Spiegel. *Pharmacology: An Introduction.* (New York: Wiley, John & Sons, 2003.).

50. For a brief account of the difference between the two terms and what they stand for, see Healy 2009, Introduction; see also http://cat.inist.fr/?aModele =afficheN&cpsidt=17461743; Ulrich Mülle *et al.* 2006.

51. Historically, their uses have been various: religious (especially in the past), spiritual heightening (to induce states of consciousness not normally experienced), recreational and medicinal. This discussion is only confined to the last mentioned.

52. See http://www.rci.rutgers.edu/~lwh/drugs/psypharm.htm; http://www. neurotransmitter. net/drugmechanisms.html; http://www.anaesthetist. com/ physiol/ basics/ receptor/receptor.htm.

53. For a brief and clear account, see http://physics.syr.edu/courses/modules/ MM/ brain/ brain.html. Nevertheless, one must bear in mind the limitations of present knowledge about the brain, given its immense complexities both at the molecular, cellular, but especially at the systems levels of its operations. For a general account of modern psychopharmaceuticals, see Spiegel 2003, Chapter 1.

54. http://www.brain.umn.edu/research/MEG.htm.

55. http://www.radiologyinfo.org/en/info.cfm?pg=fmribrain.

56. For details, see http://physics.syr.edu/courses/modules/MM/ brain/brain. html; http://en.wikipedia.org/wiki/Neuroscience; http://en.wikipedia.org/ wiki/Human_brain.

57. For details, see Healy 2009, 10-2 and Healy 2002, 198-224; http://www.neurotransmitter. net/drug mechanisms.html; http://www.anaesthetist.com/ physiol/ basics/ receptor/receptor.htm.

58. Today, science regards it as an endocrine gland, producing the hormone melatonin.

59. For details about Descartes' curious view regarding the matter, see http:// plato.stanford. edu/entries/pineal-gland/.

60. This technique is based on the discovery, in the early twentieth century, by the Russian scientist Pavlov who discovered that a dog can be trained to respond to a certain stimulus in the following way: the animal when presented with, say, a juicy chunk of meat (the unconditioned stimulus) would naturally salivate (the unconditioned response). Pavlov then rang a bell (a

neutral stimulus at the beginning of the experiment) whenever he presented the meat to the dog which then salivated in the normal way. After a few such pairings, he found that when he rang the bell even in the absence of the meat, the ringing of the bell alone (the originally neutral stimulus is now the conditioned stimulus) was sufficient to make the dog salivate (the unconditioned response). This is the general principle of classical conditioning. See http://www.learning-theories.com/classical-conditioning-pavlov.html.

61. The range of behaviour which we exhibit in our daily lives is entirely the result of our being conditioned to do so via rewards or punishment. For instance, we train a dog to pick up a stick by rewarding the animal with a dog biscuit, say, each time he does so when ordered; we whack him hard if he bites a guest to the house. We do the same to our children – good marks at school win them an expensive play station, while being rude and refusing to do assigned household chores means withholding of pocket money. See Boeree 2006; http://webspace. ship.edu/ cgboer/skinner.htm; http://psychology.about.com/od/behavioural psychology/a/ introopcond. htm.

62. For an account of its discovery, see Healy 2002, 77-101.

63. See http://www.rci.rutgers.edu/~lwh/drugs/psypharm.htm; Reynolds 1975.

64. For instance, I decide to punch you in the nose because you have just insulted me (a mental event); I deliver the punch and as a result, your nose is broken and is bleeding badly (physical event). I take a psychoactive drug (a physical event) for my depression; my depression is then lifted (mental event).

65. The blood pressure of a mother goes up (physical event) because she is worried that her son may get killed at the front; news then arrive to say that her son is on his way home as the war is now officially over; she feels greatly relieved (mental event). Her blood pressure returns to normal (physical event). Such an outcome is not compatible with epiphenomenalism.

66. For further brief details about interactionism and epiphenomenalism, see http://www. Philosophyofmind.info/mindbodyinteraction.html; for discussion of epiphenomenalism, see William Robinson, "Epiphenomenalism", http://plato.stanford.edu/entries/epiphenomenalism/.

67. This phenomenon obviously has ethical implications for clinical practice, but this aspect will not be examined here. For some discussion, see O'Leary and Borkovec 1978; Evans 2004, Chapter 9.

It has a negative counterpart, called the "nocebo" effect, an effect which consists of making the patient worse instead of better. The discussion here will ignore it as it throws up no theoretical issues not already raised by its positive counterpart.

68. Initially, the term was not used in a medical but religious context. In the Middle Ages, the faithful paid the priests to say prayers such as vespers for their dead relatives. There then arose the meaning of the term to refer to consoling words, even when perhaps insincerely uttered as those who uttered them were paid to do so. In the fourteenth century, Chaucer used it to refer to flatterers "who are the devil's chaterlaines for ever singing placebo." Frances Bacon, in the seventeenth century, similarly referred to flatterers as those who "sing a song of placebo". By the eighteenth century, the term had been incorporated into the medical context, to refer to fake remedies – see Moerman 2002; Evans 2004.

69. Thomas Jefferson in 1807 wrote to a physician friend (as cited in Brody 1980, 97): "… if the appearance of doing something be necessary to keep alive the hope & spirits of the patient, it should be of the most innocent character. One of the most successful physicians I have ever known, has assured me, that he used more bread pills, drops of colored water, and powders of hickory ashes, than of all other medicines put together." The editorial of the *British Journal of Medicine* in 1952 estimated that forty per cent of patients visiting general practitioners in England were prescribed placebos – see Brody 1980, 99.

70. One study conducted in 1985 shows that placebo is 56 per cent as effective as morphine. In another study published in *Nature*, 1984, "placebo was as powerful as a hidden injection of 8mg morphine intravenously" – cited in Koshi and Short 2007, 6.

71. Note that in 1954, an article appeared in *The Lancet* calling the placebo "a humble humbug." This shows that Beecher's new found interest was not shared by all within the Western medical establishment. The article continued: "…for some unintelligent or inadequate patients life is made easier by a bottle of medicine to comfort their ego; that to refuse a placebo to a dying incurable patient may simply be cruel; and that to decline to humour an elderly 'chronic' brought up on the bottle is hardly within the bounds of possibility" – cited in Brody 1980, 2.

72. X and Y are two events; should Y be observed to follow X, one could be tempted to infer that X and Y are causally related, that X is the cause and Y its effect. In reality, Y may not be an effect of X – thunder always follows lightening, but the lightening is not the cause of the thunder.

73. However, the notion of the no-treatment group is not itself without problems – see Moerman (2002, 26–7) who argues that it is neither logically nor conceptually possible. In one historical case cited, Moerman shows that it has violated ethical norms which today at least are expected to govern experiments involving patients.

74. This figure, although repeatedly cited since Beecher's publications on the subject, turns out to be inaccurate. In a recent review of 75 randomized placebo-controlled trials, Walsh et al. 2002 found that the proportion of patients suffering from depression who responded to placebo ranged from 12.5 per cent to 51.8 per cent. "There was an association between the year of publication and the response rate. The rate of placebo responders was shown to increase significantly in recent years' publications. Liberman [1964] found that 56% of subjects responded to the placebo treatment, whereas Petrovic [2002] found that nearly 100% of population responded to placebo. In fact, most people have experienced placebo at one time or another, suggesting that we all have the potential to develop a placebo response" – cited in Koshi and Short 2007, 9–10.

75. In 2001, two medical researchers, Hrobjartsson and Gotzsche, in Copenhagen published a meta-analysis of 130 placebo trials which included a control no-treatment group. They concluded that there is little evidence that placebos had any significant clinical effects. However, as pointed out by Evans (2004, 28) when their work is in turn subjected to critical scrutiny, it has been found to be wanting: "What promised to be the final, definitive word on

placebos turned out not to be a proper study, full of flaws, and capped by an inaccurate summary."

76. For details about specific trials, the range of the trials, the types of medical interventions, see Moerman 2002; Evans 2004; Koshi and Short 2007; Niemi 2009.

77. Unlike some other forms of treatment, placebo surgical procedures though not easy to mount could be done, in cases where the patients are given a general anaesthetic.

78. See Moerman 2002, 9–10; Niemi, 2009 – healing of duodenal cancer by placebo effect was 36.2 to 44.2 per cent of the 3325 patients involved in 79 studies.

79. See Niemi 2009, about the extraordinary fate of one patient who, indeed, was cured by the placebo effect, but later killed by having read that the new anti-cancer drug (Krebiozen) he had been given was really worthless – this new belief destroyed the placebo effect responsible for the cure, and he died of the cancer of the lymph nodes with which he had originally been diagnosed.

80. One interesting thing follows: apparently in so-called double-blind tests, patients as well as the medical staff involved are pretty good at guessing who are taking the experimental drug and who the placebo. If patients guess that they are taking a placebo, the placebo effect is diminished. Similarly, if they guess that they are taking the experimental drug, the placebo effect may be enhanced. (Researchers have found that in 23 out of 26 cases where they checked on this point, both patients and staff did better than chance at guessing who got what.) In other words, given the possibility of such an enhanced placebo effect, "the ratio of the specific effect of the real drug to its placebo component becomes even smaller still." On these points, see Evans 2004, 40.

81. See Koshi and Short 2007.

82. Ibid., 14.

83. For some of the mechanisms, see Koshi and Short 2007, 6–9; Niemi 2009.

84. Evans (2004, 44–69; 134–5) seems to place a lot of emphasis on the mechanisms governing the acute phase response of the immune system. He also focuses on conditioning, via the experience of humans as well as non-human animals; such experience leads to expectancy on the part of the agent involved, whether human or non-human. In other words, the principles of conditioning are applicable to both the human and animal contexts and can account for expectancy in either. Their immune systems can also be conditioned to display the placebo effect – see, Evans 2004, 81–6, 99–103. However, Evans is not denying that in the case of humans, there is a significant additional dimension to the placebo effect based on their ability to use language; this more complex cognitive ability means that for humans, belief/experience/expectancy are consciously and linguistically mediated.

    For an early documentation of the role played by expectancy in the phenomenon of the placebo effect (in the human context), see Shapiro 1971.

85. For a critique of the inadequacy of the reductionist approach in the context of discussing psychosomatic medicine and the placebo effect, see Foss 1998.

86. For an instance, see Evans 2004, Chapter 3. (However, to determine Evan's own view, this chapter should be read in conjunction with his Chapter 4. Together they seem to support the assessment that Evans adheres to eliminative materialism, rather than crude materialism.)

Under eliminative materialism, statements such as "I have a pain" should be replaced by "My C-fibers are firing", as the latter is correct scientific talk, whereas the former is mentalistic talk, which refers to something which is plainly subjective, and therefore, not objectively checkable or measurable. As a result, it would eventually drop out altogether, and people will find it natural to use scientific talk in common conversation. Even should some continue to use mentalistic talk, the rest would know that the speakers are really talking about specific events which are taking place in the brain.

87. Pavlovian conditioning is a crude form while Skinner's operant conditioning is a more sophisticated form of behaviourism. For a brief but succinct critique of the various philosophical positions regarding the mind/body problem, see Brody 1980, Chapters 3 and 4.

88. See, for instance, Niemi 2009.

89. Moerman (2002, 90–1) says: "... quite convincing evidence that all forms of psychotherapy (psychoanalytic, Adlerian, Eclectic, Client-centered Rogerian, Behaviourist, etc) are effective and they are all equally effective. The typical therapy client is better off than 75% of untreated individuals." On this particular point, Evans (2004, Chapter 8) agrees with Moerman.

90. Evans (2004, 93–94), who does not subscribe to the Moerman explanatory model, is keen to rule that "positive thinking" is beyond the scientific pale and that study of the placebo effect should be withheld from such a domain.

91. Moerman writes from the standpoint of anthropology; he says nothing explicitly about philosophy. Instead, he concentrates on meaning (in beliefs), in the main, from an empirical standpoint. Hence, he appears to privilege mind over body – this author would owe him an apology should his account have no truck with dualism in any form. He (2002, 154) has a few brief remarks, especially towards the end of his book, which has some philosophical import: "People are simultaneously biological and cultural creatures. Biology and culture interact, and are equal partners in who and what we are."

If this kind of remark is taken into account, then Moerman cannot be taken to privilege mind over body, and his position would then be near identical – if not identical – to the third approach, the one which transcends Cartesian dualism. If this were indeed so, then he could be in agreement with Brody's explicit philosophical stance in arguing that the notion of person is primitive. However, his text (2002), as it stands, gives very little guidance about this matter. It is true that he rejects privileging body over mind: He (2002, 137) writes: "Many people see the human body as a machine...If the body is a machine, then we might well be surprised each time we find people responding to inert medications...In the mechanistic tradition that still underlies much of modern biomedicine, believing in the power of a placebo to erase pain is as irrational as filling the gas tank of your car with Earl Grey."

92. Our notion of moral/legal responsibility rests on the primitive concept of personhood. In the law of homicide, before a defendant can properly be convicted, the court must satisfy itself on two counts. First, it must be established beyond reasonable doubt that indeed the person who has committed the murder is the person standing in the dock – to satisfy this requirement, one can use today's technology to imagine a CCTV camera recording every move and action of the person committing the act of killing the victim. This requirement is about locating the body of the person, so to speak, to make sure that there is no mistaken identity. Second, however, such a condition on its own is not sufficient to secure conviction – the court must also be satisfied that the defendant is *compos mentis*, that is, of sound mind and that he had the intention to kill. The two requirements must be jointly satisfied to warrant conviction. That is why confession of a crime on its own is not sufficient to warrant conviction – the court must be satisfied that the confessor is indeed the person who committed the act (the sense of "act" in this context would be equivalent to catching the individual in the very physical act of wielding the knife, and so on.)

    Occasionally, such a theory of personality breaks down in a particular application. Suppose X committed a murder. Following the event, he was not immediately apprehended by the police. He then suffered amnesia not merely for the murder but also who he was; he drifted away from his home town, say, even emigrated. He then started an entirely new life in the new country with a new identity. He lived a peaceful and fruitful life as a citizen for forty years before Interpol finally caught up with him. A court would have difficulty convicting such a person, as the two requirements cannot be jointly satisfied at the time of the trial. His amnesia concerning his former identity and criminal action (provided the court is convinced it is a genuine case) would mean that he is no longer the same person as the person who committed the act of murder 40 years ago. Continuity of memory is a criterion of the identity of a person; crucial/key loss of memory disrupts that identity. (This loss should not be equated with normal loss of day-to-day memory, say, about what one has eaten or worn on a certain date, and so on.)

93. It is an account derived from the writings of Strawson 1959 and other like-minded philosophers. For a brief but succinct account, see Brody 1980, Chapter 5.

94. The German Medical Association (BÄK) in March 2011 issued a report which endorses the extensive use of placebos by German doctors, while admitting that medical science today cannot fully explain why and how they work. See Abby D'Arcy Hughes at http://www. guardian.co. uk/science/2011/mar/06/half-german-doctors-prescribe-placebos; "Mind-altering" Report endorses German GPs' use of placebos, in *The Guardian*, Monday 7 March 2011; "Placebos can reduce side effects and costs, argue German doctors" http://www.bioethics.ac.uk/news/ placebos-can-reduce-side-effects-and-costs-argue-german-doctors.php; Bundesärztekammer, "Placebo in der Medizin – Mehr als nur einbildung", Berlin: 02/03/2011 at http://www. bundesaerztekammer.de/page.asp?his=3.71.8899.9061.9064.

9 Nosology: The Monogenic Conception of Disease

1. Different writers have their own preferred schemes – see for instance, Wulff and Gotzsche 2000; Ronald David at http://www.nosology.net/approach. php.
2. For a detailed account of his method, see Faber 1978, Part I.
3. However, one should not ignore some earlier very impressive discoveries in the field, between 1835 and 1889 – see table 1.2 in Evans 1993, 9.
4. We shall examine these postulates in some detail later.
5. It is claimed by some scholars that there have been two separate discoveries of the bacterium which causes cholera – see, for instance, http://www. ph.ucla.edu/epi/snow/firstdiscoveredcholera.html.
6. One well-known one is the Gram stain, which distinguishes the Gram positive from the Gram negative type of bacteria. The former type of bacterium is more sensitive, for instance, to the antibacterial action of penicillin, iodine and acids. For a brief account of the technique, see http://www. microbiologyprocedure.com/staining-methods-in-microbiology/differential-stains. html.
7. For a short account, see http://www.sciencemuseum.org.uk/broughttolife/ techniques/miasmatheory. Aspx. It arose primarily during the Middle Ages and may be considered as a theory of disease origin based on environmental factors. Miasma was a poisonous vapour, in which were suspended particles of decaying matter, giving forth a foul smell.
8. For an account (critical), see http://209.85.229.132/search?q=cache: LyTIc2gIhLgJ:www.teachnet-uk.org.uk/2007%2520Projects/Hist-Medicine_ Four_Humours/Four_Humours_Theory/ Theory%2520of%2520the%2520F our%2520Humours%2520in%2520the%2520History%2520powerpoint.pp t+humour+theory+of+disease&cd=9&hl=en&ct=clnk&gl=uk.
9. For a brief discussion, see http://www.olemiss.edu/courses/phcy201/cdh2. htm ; http://schaechter.asmblog.org/schaechter/2008/03/plentisillin-an. html ; Brown 2004; Bud, 2007.
10. For one significant dissenting voice about this "Whig" account of the history of modern medical therapies, see McKeown 1979. Also some serious disenchantment with antibiotics began slowly to set in for a variety of reasons, such as (a) large quantities have been known to be prescribed, which, *ex hypothesi*, eliminate all bacteria in the guts, whether harmful or non-harmful, (b) the ability of bacteria to replicate and mutate quickly leads to strains which are resistant to extant antibiotics, creating what is sometimes called "superbugs" – see http://www.iom.edu/~/media/Files/ Activity% 20Files/Public Health/MicrobialThreats/Davies.ashx.
11. http://nobelprize.org/nobel_prizes/medicine/laureates/2005/.
12. More commonly, this is called pre-frontal leucotomy, which turned out to be a very controversial brain surgery. So did the removal of the thyroid gland, as noted in an earlier chapter.
13. Cited by Carter 2003, 1.
14. However, genetic factors are internal to the (human) organism, unlike poisons, coal/asbestos dust or infective agents which are external in provenance.

15. See Lakatos 1970, 118–9.
16. As quoted by Evans 1993, 20. The two points made so far about the aetio-logical/monogenic conception of disease show clearly that it relies upon a particular view of cause, the details of which will be explored not in this but in a later section of this chapter and in the rest of the book. We shall also see that critics do exist who claim that the research programme has run into some very serious anomalies.
17. This raises the key issue, which will be examined in detail later in the chap-ter, of who count as disease bearers, that is to say, whether the criterion implied is simply that the bacteria associated with the disease be present in the person, or that the person should also exhibit the symptoms/signs associated with the bacteria present in the body.
18. See R.J.W. Rees 1988, "Leprosy" at http://bmb.oxfordjournals.org/cgi/reprint/44/3/650.pdf.
19. See Waller 2004; http://www.virology.ws/2010/01/22/kochs-postulates-in-the-21st-century/.
20. See Evans 1993, Chapter 7 for an assessment.
21. See Thagard 2000, 47–8.
22. However, when made public, the discovery of the actual bacillus became a surprise as, up to then, the consensus in the medical world held that the stomach is sterile and, therefore, could not harbour any bacteria. This puz-zle was later removed when it was found that this particular kind of bacteria has two abilities: to burrow beneath the mucous layer in the stomach as well as to produce an enzyme, urease, which can turn the urea in the gastric juice into ammonia, which, as it is alkaline, can neutralize the acid in the stomach – see Thagard 2000, 58.
23. He later abandoned the experiment; in any case, his piglets were growing too big for easy keeping and so he had them put down. However, he did not publicize this failure but kept it to himself. See Marshall's Nobel Lecture: http://nobelprize.org/nobel_prizes/medicine/laureates/2005/marshall-lecture.pdf.
24. See Philip Shayne and Wendi S. Miller, 10 November 2010, "Gastritis and Peptic Ulcer Disease" at http://emedicine.medscape.com/article/776460-overview. According to the authors: "Gastritis technically refers to endo-scopic or histological findings of inflammatory changes in the gastric mucosa; however, the term is often used clinically to refer to the symp-toms of dyspepsia. The most common causes are *Helicobacter pylori* bacterial infection and the use of nonsteroidal anti-inflammatory drugs (NSAIDs) and aspirin..."; "Peptic ulcer disease (PUD) refers to the development of a discrete mucosal defect in the portions of the gastrointestinal tract (gastric or duodenal) exposed to acid and pepsin secretion. Presentations of gastritis and PUD usually are indistinguishable... and thus the management is gener-ally the same."
25. See Thagard 2000, 59.
26. For instance, see Thagard 2000, 59–61; Evans 1993, Chapter 3.
27. One could just as easily use the genetic instead of the infectious-agent model of disease causation to illustrate the same philosophical/methodo-logical points which follow in this chapter and the rest of the book. For one account of such complexities, see Johnjoe McFadden, "ADHD's roots are

complex", *The Guardian,* 01/10/10 at http://www.guardian.co.uk/ comment-isfree/ 2010/sep/30/attention-deficit-disorder-genetic-roots.

28. For some accounts, see Andrew Brennan 2003 at http://plato.stanford.edu/ entries/necessary-sufficient/; Norman Swartz 1997 at http://www.sfu.ca/philosophy/swartz/ conditions1.htm; Stephen Downes Moncton 2008 at http:// halfanhour.blogspot.com/2008/01/necessary-and-sufficient-conditions. html.

29. See Carter 2003, Chapter 2 and 203, Note 2; Susser 1973, 6.

30. However, oxygen, though necessary, is not also a sufficient condition as its presence alone does not guarantee that a fire would occur, as other conditions must also be present at the same time, such as inflammable material, a source of fire (a lit match, lightning, and so on), no-one putting out the fire just as it started, no wind coming from a suitable direction and at a speed to fan and spread the conflagration, and so on. From even such a brief account, one would immediately grasp that neither necessary nor sufficient conditions can, in principle, be exhaustively listed. The chain can be stretched indefinitely further and further back, as illustrated in the common nursery rhyme:

> For want of a nail the shoe was lost.
> For want of a shoe the horse was lost.
> For want of a horse the rider was lost.
> For want of a rider the battle was lost.
> For want of a battle the kingdom was lost.
> And all for the want of a horseshoe nail.

This reminds us that in causal investigation, in seeking for an explanation in terms of necessary and sufficient conditions, the remit is context-dependent. For instance, in trying to explain the fortunes of Napoleon, it is not relevant to invoke the particular marriages of his grandparents or the conceptions of his parents or indeed, even his own conception, although all these are events which qualify to be necessary conditions for Napoleon's numerous military successes as well as critical defeats. We shall return to this point later.

31. This list is culled from http://hcd2.bupa.co.uk/fact_sheets/html/peptic_ulcer.html ; http://www.cdc. gov/ulcer/md.htm.

32. See http://www.nutramed.com/digestion/ulcer.htm.

33. See http://nobelprize.org/nobel_prizes/medicine/laureates/2005/marshall-lecture.pdf.

34. See, for instance, http://digestive.niddk.nih.gov/ddiseases/pubs/nsaids/; http://www. netdoctor.co.uk/ diseases/facts/pepticulcer.htm ; http://www. emedicine. com /EMERG/ topic820.htm# section~ medication.

35. http://nobelprize.org/nobel_prizes/medicine/laureates/2005/marshall-lecture.pdf.

36. For instance, Thagard 2000, by and large, falls into this category. His term "principal" may be understood in two ways: (a) statistically – in the majority of PUD cases, *H. pylori* can be found and is the cause of PUD; (b) *H. pylori* exists as one of a set of causal factors but as eliminating it eliminates PUD in the majority of cases, it can be said to be the principal cause. (We shall be examining this second understanding in Chapter 11.)

37. See Marshall's Nobel Lecture at http://nobelprize.org/nobel_prizes/medicine/ laureates/2005/ marshall-lecture.pdf.

38. However, this more restricted claim, on its own, may not have generated enough attention and excitement to attract the award of a Nobel Prize, at least in the opinion of this author.
39.

Primary tuberculosis	Secondary tuberculosis
No symptoms	Fever and weight loss
Minor self-limiting illness	Enlarged lymph glands
Grumbling illness with fevers	Persistent cough
Overwhelming illness such as tuberculous meningitis	Skin rashes
	Septicaemia (miliary tuberculosis)

[The above is from Bhopal 2008, 171.]

40. This observation should not be distorted to mean that causal discoveries are in reality no more than tautological truths. Causal claims are factual matters; however, when the factual discovery is confronted by counter-examples or anomalies, the defenders of the claim may adopt the strategy of guaranteeing its truth by taking the definitional turn.
41. Similarly, he claims that retroviruses are neither necessary nor sufficient for producing cancer.
42. However, see Jon Cohen, "Fulfilling Koch's postulates", *Science*, 266 (2 December 1994); **http://www.sciencemag.org/feature/data/cohen/266–5191-1647a.pdf**. Duesberg's critics claimed that in 1993 evidence had emerged which would satisfy Koch's postulates. To this, Duesberg threw down another challenge, namely, that an epidemiological study involving two groups with the same age, lifestyle (no drugs) but with different HIV status be conducted – if AIDS-defining diseases were found to be significantly more in the HIV positive group than the HIV negative group, he would concede that HIV is a likely cause of AIDS.
43. See also http://chestofbooks.com/health/natural-cure/Ross-Horne/Health-and-Survival-in-the-21st-Century/HIV-AIDS-Addendum-A-Conversation-With-Peter-Duesberg.html. See Evans 1993, 174; Thagard 2000, 61.
44. See Evans 1993, 64 and 137–38.
45. See Cohen cited in Note 42 for evidence that Duesberg is willing to give up these postulates.

10   Linear Causality and the Monogenic Conception of Disease

1. See Lakatos 1970.
2. There is an alternative conception in modern medicine which will be explored in Chapter 12.

3. For an accessible and succinct account of cause in general and of the Humean account in particular, see Hanfling 1980.
4. See, for example, http://kidshealth.org/parent/medical/genetic/down_syndrome.html.
5. "Autosomal" means that the condition does not involve sex chromosomes.
6. See Bhopal 2008, 134.
7. Indeed, PKU involves a single genetic defect which leads to a specific metabolic disorder. However, this should not lead to the oversight that the genetic defect constitutes only a necessary, but not also sufficient, condition of the manifestation of the specific metabolic disorder in the bearer of the genetic condition. In this, it differs from *H. pylori* in that regarding the latter, the bacterium is neither a necessary nor a sufficient condition for PUD.
8. In what follows a little later, we shall be looking at some criteria in terms of which in some contexts one would single out as a single factor from the combination of factors for special emphasis calling it "the cause". However, such a context is not primarily an explanatory/scientific context of enquiry.
9. See, for instance, http://www.ccohs.ca/oshanswers/chemicals/synergism.html.
10. See Lee 1989b.
11. Since the beginning of the twentieth century, mathematical logic understands propositions such as "two plus two equal four" (p) to be analytical/definitional/tautological truths. The truth of p depends entirely on the way in which "2", "4", "+", "equals" have been defined.
12. See Mackie 1974; Wulff 1984.
13. Removal of the spleen is not only associated with an increased incidence of pneumococcal infection but also with other forms of sepsis, for instance, Lyme disease.
14. See Hart and Honoré 1959.

11   Determining "the Cause": Controllability and Random Controlled Trials

1. See http://www.niaid.nih.gov/topics/commoncold/Pages/default.aspx.
2. For details, see Thagard 2000, 56–64.
3. See "Helicobacter pylori in peptic ulcer disease NIH Consensus Statement", 1994 Jan 7–9; 12(1) 1–23 at http://consensus.nih.gov/1994/1994HelicobacterPyloriUlcer094html.htm.
4. For a brief account, see R. Johnson 2008, section 4 of Kant's Moral Philosophy at http:// plato.stanford.edu/entries/kant-moral/.
5. For a recent critical assessment, see Ralph Wedgewood 2009 at http://users.ox.ac.uk/~mert1230/ Instrumental%20Rationality.a4.pdf.
6. Collingwood (1938, 97–8) writes that in this sense, "a cause is necessary (a) in its existence, as existing whether or no human beings want it to exist (b) in its operation, as producing its effects no matter what else exists or does not exist... The cause leads to its effect by itself, or 'unconditionally'"; "in other words the relation between cause and effect is one-one relation ...."
7. In the light of the discussion which follows, as well as the arguments set out in Chapter 12, this appears ironic as Bradford Hill was the leading epidemiologist of that period.

8. See Bradford Hill 1965; Lock 1994; Silverman 1985; Cochrane 1999; Wulff and Gotzsche 2000, especially Chapter 6.
9. See Silverman 1985.
10. See http://inventors.about.com/library/inventors/blantisceptics.htm.
11. The Cochrane Collaboration was set up in the wake of A. L. Cochrane's article, entitled *"Effectiveness and Efficiency: Random reflections on health services"*, first published in 1972. Its remit is to conduct systematic reviews of the effects of healthcare interventions as recorded in published trials, thereby contributing to what is now called evidence-based medicine. See http://www.cochrane-collaboration.com/.
12. See Evans, Thornton and Chalmers 2007 at http://www.jameslindlibrary. org/pdf/testing-treatments. pdf.
13. See Cochrane, 1972/1999, 22.
14. The operative word "seemingly" is this author's own gloss. It is to draw attention to the fact that as no RCT would or could be expected for such products, there is no evidence either way as to whether such small variations in their chemical composition would produce different effects from their prototypes for which RCT had been conducted. This default axiom, undoubtedly, is economically and practically sound, but is it also methodologically sound? There may be room for some doubt regarding the latter aspect.
15. See Healy 2009, Chapter 28; Silverman 1985, 151; see also BBC2 2010 at http://www.bbc. co.uk/programmes/b00q9jfs#p00636b7 – Dr Patrick Vallance, Head of Drug Discovery of GSK is reported as saying that (a) clinical trials involve at best thousands of cases, but the true side effects require hundreds of thousands of cases; (b) therefore no drug is totally safe and uncertainty is built into the matter.
16. For further details, see http://yellowcard.mhra.gov.uk/.
17. Cochrane (1972/1999, 25) writes: "The main job of medical administrators is to make choices between alternatives. To enable them to make the correct choices they must have accurate comparable data about the benefit and cost of the alternatives. These can really only be obtained by an adequately costed RCT."
18. See http://www.nice.org.uk/ ; http://www.bmj.com/cgi/content/full/328/7439/536.
19. See Healy 2009, Chapter 28.
20. For one account presented via maps, see http://www.ph.ucla.edu/epi/snow. html. Andrew Hayward's lecture on the subject may be found at: http://www.pitt.edu/~super1/ lecture/lec1151/index.htm.
21. For one account, see http://en.wikipedia.org/wiki/Experiment.
22. We have used the term "verified", as it seems a very "natural" term to use here; however, the usage here is meant in a neutral way without any intention of either endorsing Popper's philosophy of science or challenging it, as this would be beyond the remit of this limited discussion. For the author's own critical assessment of Popper's methodology, see Lee 1985.
23. We shall be exploring this difference further between the two contexts – drug trial and epidemiology – in the next chapter.
24. An electronic version may be found at http://www.gutenberg.org/etext/27942.

25. For an overall account of Mill's thoughts, see F. Wilson , 2007 at http://plato. stanford.edu/entries/ mill/ ; for one account of Mill's canons for medicine and biology, see Schaffner 1993, Chapter 4.6, 142–46 which can also be found at http://books.google.co.uk/books?id= aVyIW cRmne8C &pg=PA143&lpgP A143&dq=Mill's+experimental+methods&source=bl&ots=cOTzK5atAJ&sig =lv2ORXsS77RN4ZD8S08marmtACw&hl=en&ei=_H3ISbXH43w0gT82LD KDA&sa=X&oi=book_result&ct =result&resnum =5&ved= 0CBQQ6AEwB DgU#v=onepage&q=Mill's%20experimental% 20 methods&f=false.

26. This is confounding; the term may be defined as: "the distortion of the measure of an association by other (confounding) factors that influence the outcome and risk factor under study" – see Bhopal 2008, xxviii.

27. http://lhc.web.cern.ch/lhc/.

28. http://www.scientificamerican.com/article.cfm?id=what-exactly-is-the-higgs.

12   Epidemiology: "Cinderella" Status?
What Kind of Science is it Really?

1. This does not mean historically that there were not earlier attempts, such as in Mexico, the Vatican, and more recently in Germany of the Third Reich.

2. See http://www.ncbi.nlm.nih.gov/pmc/articles/PMC2038856/. However, as pointed out by the authors themselves, German scientists, in 1939, had already raised such a link but probably because of the war and political reasons, nothing came of this earlier publication. Other small-scale investigations after WWII also confirmed it. In 1950, the results of a larger-scale study in the United States (involving 601 patients) strengthened the claim even further. Doll's and Bradford Hill's case control study, conceived and begun in 1947, involved 20 London hospitals; the lung-carcinoma group as well as the control group (which consisted of patients with diseases other than cancer) each numbered 709 participants, with a male/female gender break down as 609/60 in each group. In the former group, the figure of non-smokers among the men was 0.3 per cent and 31.7 per cent among the women; in the latter group, men non-smokers were 4.2 per cent and women non-smokers were 53.3 per cent. It followed that a significant and clear relationship between smoking and carcinoma of the lung emerged, no matter which measure of tobacco smoking was used. This much larger-scale study, consequently, ruled out exposure to tarmac or car fumes, but not tobacco smoking, as being implicated in lung carcinoma.

3. On the differences between case control and cohort studies (as well as other studies used in epidemiological research), see, for instance, Bhopal 2008, 15.

4. The first results were reported in *BMJ* 1954(228) 1451–5 by Doll and Hill, "The mortality of doctors in relation to their smoking habits: a preliminary report." This was followed in 1976 by Doll and (Richard) Peto (in *BMJ* 1976(2) 1525–36), "Mortality in relation to smoking: 20 years' observations on male British doctors." The latest update can be found in *Br J of Cancer* 2005(92), Doll, Peto, J. Boreham and I Sunderland's "Mortality from cancer: 50 years' observations on British doctors." See also P. Boyle, "Tobacco smoking and the British doctors' cohort" in *Br J of Cancer* 2005(92) 419–20.

5. See Richard Peto 2005, "Nature, Nuture and Luck" (a celebration of Doll's life-time achievements) – http://www.ctsu.ox.ac.uk/news/nature-nurture-luck-oxford-today-article.pdf/view.

6. On these points above, see Kuhn 1970 (For one brief account of Kuhn's work including his notion of paradigm, see http://plato.stanford.edu/entries/thomas-kuhn/ ; Lakatos and Musgrave 1970.)

7. This is not helped by the fact that Kuhn was often not too careful in the way in which he himself used the term. See Lakatos and Musgrave 1970.

8. For an account of what the science looks like, see Bhopal 2008.

9. See Bhopal 2008, 131.

10. This disjunct is added to Bhopal's account by this author.

11. See, for instance, http://www.emedicine.com/EMERG/topic820.htm#section ~medication.

12. One may proffer the following definition: that branch of biology which studies the relations and interactions between organisms and their abiotic environments as well as among the organisms themselves, at the level of communities, populations, ecosystems both at local and global scales. For a brief account, see Barbara Stewart 2008, "The Science of Ecology" at http://geologyecology. suite101.com/ article.cfm/the_science_of_ecology. For a detailed account, see McIntosh 1985. For a quick account of Odum's life and work, see http://www.chattoogariver.org/ index. php? req= odum&quart=F2002.

13. Relativism maintains that Truth does not exist, but only truths, each relative to its own historical period and age, its own social context. To take morality as an instance – relative to traditional Eskimo society, parricide was morally right; relative to Catholic beliefs, contraceptives are immoral, while according to contemporary liberal views, contraceptives constitute a great advance and benefit, not only to womankind but to all humankind. Furthermore, opposing truths are not commensurable. In the scientific domain, Kuhn's account, it is said, claims that theories embedded in different paradigms are incommensurable – one cannot compare the Ptolemaic earth-centred theory with the Copernican heliocentric theory in astronomy to determine which is superior or is a better fit to "reality." Indeed, there is no Reality, as each theory/paradigm lays down its own reality. For a critique of the thesis, see Lee 1984.

14. We introduce the term here in a technical sense, namely, any entity or process which is capable of effecting change in any other entity or process.

15. These three may also be classified into two major groups: biotic agents (plant or animal) on the one hand, and abiotic (physical/chemical) agents on the other.

16. See Lee 1999.

17. The phenomenon is also called global warming. These terms can be understood in one of two ways: that: (a) Earth's temperature has risen over the last hundred years or more; (b) this increase is anthropogenic. Global warming sceptics may be divided into groups – those who deny (b) but accept (a), and those who deny (a), thereby also denying (b) as an implication of (a).

18. Greenhouse gases have always existed in Earth's atmosphere, keeping it roughly at 33 degrees Celsius warmer than it would be without them. In their total absence, organisms would have not have evolved – organisms are carbon-based and would not normally thrive in the absence of warmth and

light which come from the Sun. Greenhouse gases serve the useful function of retaining some of that radiation from the Sun while remitting the rest. It is their excess in the atmosphere which poses a threat. It is now acknowledged that global warming is alarming for the future of both humans and non-humans.

19. For one account regarding the relationship between genetic endowment and intelligence, see Dickens and Flynn 2001.
20. For an earlier version, see Lee 2010.
21. This has led some to propose that vaccination programmes against infectious diseases should be made compulsory, on grounds of fairness, so that costs as well as benefits should be borne by all.
22. Just one instance of recent research to show the "ecosystem" effects of genes and their manifestations in ecological studies, see Joshua Mutic and Jason Wolf 2007, "Indirect genetic effects from ecological interactions in Arabidopsis thaliana", *Molecular Ecology* 16(11) 2371–81; http://www.ls.manchester.ac.uk/research/publications/archive/article/?id=17561898.
23. The vaccine is part of clinical medicine; however, it is also part of preventive medicine.
24. Admittedly, it would be difficult to satisfy the requirement of double-blinding, unlike an experiment in clinical research which involves the participants swallowing pills.
25. See Bhopal 2008, 15, the * note to table 1.5.
26. Recall the episode about the results of the research of Japanese science during World War II in Chapter 6, Note 9 – the United States suppressed the atrocities committed by the Japanese researchers on prisoners-of-war in their experiments in exchange for taking over the fruit of their labour, on the grounds that such data could never be obtained under more civilized conditions of doing science.
27. Perhaps an encouraging sign of the time is that there has appeared a shift from the atomistic model to the ecosystemic model of understanding infectious diseases. For instance, an account of gastritis/PUD within the latter methodological framework may be found in Shayne and Miller, 10 November 2009, cited earlier in Chapter 9 at http://emedicine.medscape.com/article/776460-overview. The authors write: "The pathogenesis of peptic ulcer disease is multifactorial and results from an imbalance of the aggressive gastric luminal factors, acid and pepsin, and defensive mucosal barrier functions of mucus and bicarbonate. Several environmental and host factors contribute to ulcer formation by increasing gastric acid secretion or weakening the mucosal barrier. Environmental factors include NSAID use, cigarette smoking, excessive alcohol intake, and extreme emotional or physical stress. Host factors include *H. pylori* and other infections as well as hypersecretory states such as Zollinger-Ellison syndrome."

Conclusion

1. This status has been bestowed on it by leading medical teaching institutions world-wide such as universities, with established chairs in the subject.
2. Judged by the standards of a dog show, those of a cat show may be deemed to be "sub-standard". However, this deeming is neither here nor there, as

cats are not dogs. Leaning on Aristotle, one can say that different activities pursue their own excellences each in the way the nature of its subject matter permits; that one must not, therefore, ask of a subject matter a degree, say, of certainty/precision of which it is inherently incapable. Epidemiological research must work within a more complex causal framework, and is subject to ethical, not to mention also practical constraints which do not permit experiments to be set up exposing individuals to agents the researchers have reason to believe are harmful.

References and Selected Bibliography

Ackerknecht, E.H. *Therapeutics from the Primitives to the Twentieth Century*. (New York: Hafner Press, 1973).

Albury, W.R. 'Ideas of Life and Death', in Bynum and Porter, Vol. 1 (London: Routledge, 1993).

Allen, Garland E. *Life Science in the Twentieth Century* (Cambridge: Cambridge University Press, 1979).

Amsterdamska, O. and Anja Hiddinga. 'The Analyzed Body', in Cooter & Pickstone (2003).

Aronowitz, Robert A. *Making Sense of Illness: Science, Society, and Disease*. (Cambridge: Cambridge University Press, 1998).

Bala, Arun. *The Dialogue of Civilizations in the Birth of Modern Science*. (New York: Palgrave Macmillan, 2006).

Baldry, Peter, *The Battle Against Bacteria: A Fresh Look*. (Cambridge: Cambridge University Press, 1976).

BBC2. Horizon's *Pill Poppers*, broadcast on 20 January 2010: http://www.bbc.co.uk/programmes/b00q9jfs#p00636b7.

Bentall, Richard. *Doctoring the Mind: Why Psychiatric Treatments Fail*. (London: Allen Lane, 2009).

Bernard, Claude. *An Introduction to the Study of Experimental Medicine*. Trans. Henry Copley Greene. (New York: Dover Publication, 1957).

———. *Claude Bernard's Revised Edition of "His Introduction A L'Etude de la Médecine Expérimentale."* Preface by Paul F. Cranefield (New York: Science History Publications/USA, 1976).

Bhopal, Raj S. *Concepts of Epidemiology: an integrated introduction to the ideas, theories, principles and methods of epidemiology*. (Oxford: Oxford University Press, 2002 & 2008).

Black, D.A.K. *The Logic of Medicine*. (Edinburgh and London: Oliver Boyd, 1968).

Blalock, H.M. *Causal Inference in Non-experimental Research*. (Chapel Hill, NC: University of Carolina Press, 1964).

Blume, Stuart. 'Medicine, Technology and Industry', in Cooter & Pickstone (2003).

Booth, Christopher C. 'Clinical research since 1945', in Lawrence (1994).

Boeree, C. George. 'B.F. Skinner', 2006. http://webspace.ship.edu/ cgboer/ skinner. htm.

Bracegirdle, Brian. 'The Microscopical Tradition', in Bynum and Porter, Vol. 1 (London: Routledge, 1993).

Bradford Hill, Austen. 'The Environment and Disease: Association or Causation?' *Proceedings of the Royal Society of Medicine*, 58 (1965): 195–300.

———. For summary of criteria of cause: http://courseweb.edteched.uottawa.ca/ EPI6181/ Course_ Outline/Causes.htm.

Brandt, Allan M. and Martha Gardner. 'The Golden Age of Medicine?' in Cooter & Pickstone (2003).

Brock, W.H. 'The Biochemical Tradition', in Bynum & Porter, Vol. 1 (London: Routledge, 1993).

Brody, H. *Placebo and the Philosophy of Medicine*. (Chicago: University of Chicago Press, 1980).

Brown, Kevin. *Penicillin Man: Alexander Fleming and the Antibiotic Revolution*. (Stroud: Sutton, 2004).

Bud, Robert. *Penicillin: Triumph and Tragedy*. (Oxford: Oxford University Press, 2007).

Burger, Alfred. *Burger's Medicinal Chemistry and Drug Discovery*. Fifth Edition, edited by Manfred E. Wolff. (New York and Chichester: John Wiley & Sons, Inc., 1995).

——. 'The Conceptual Background and Development of Medicinal Chemistry,' Chapter 1 in Wolff (1995).

——. *Burger's Medicinal Chemistry and Drug Discovery*. Vol. 1, Sixth Edition, edited by Donald J. Abraham et al. (New Jersey: John Wiley & Sons, Inc., 2003).

——. 'Drug Discovery: The Role of Toxicology,' Chapter 16, Vol. 2, edited Abraham *et al.* (2003).

Bynum, W.F. 'Nosology', in Porter and Bynum, Vol. 1 (London: Routledge, 1993a).

——. 'The rise of Science in Medicine, 1850–1913', in W.F. Bynum *et al.* (2006).

Bynum, W.F, Anne Hardy, Stephen Jacyna, Christopher Lawrence, E.M Tansey. *The Western Medical Tradition: 1800 to 2000*. (Cambridge: Cambridge University Press, 2006).

Bynum, W.F. and Roy Porter. Eds. *Companion Encyclopaedia of the History of Medicine, 2 vols*. (London: Routledge, 1993a).

——. *Medicine and the Five Senses*. (Cambridge: CUP, 1993b).

Cantor, David. 'Cancer', in Bynum and Porter (1993a).

——. 'The Diseased Body', in Cooter and Pickstone (2003).

Caplan, Arthur L. 'Does the Philosophy of Medicine Exist?' *Theoretical Medicine and Bioethics*, Vol. 13, No. 1 (1992).

Carter, K. Codell. *The Rise of Causal Concepts of Disease: Case Histories*. (Aldershot: Ashgate Publishing Limited, 2003).

Channell F. David. *The Vital Machine: A Study of Technology and Organic Life*. (New York: Oxford University Press, 1991).

Christofferson, Ralph E. and J. Joseph Marr. 'The Management of Drug discovery', Chapter 2, Wolff (1995).

Cochrane, A.L. *Effectiveness and Efficiency: Random reflections on Health Services*. 1st ed. 1972. (Cambridge: The Royal Society of Medicine Press Limited, 1999).

Coghlan, Andy. 'Clotted Milk', *New Scientist,* (27 September 1997) 10. [First reported in *Nature Biotechnology,* 15(1997) 971]

Collins, Francis. *The Language of Life: DNA and the Revolution in Personalised Medicine*. (New York: Profile Books, 2010).

Collins, Harry and Trevor Pinch. *Dr Golem: How to Think About Medicine*. (Chicago: Chicago University Press, 2005).

Collingwood, R.G. 'The So Called Idea of Causation', *Proceedings of the Aristotelian Society*, Volume 85, 1938.

Comte, Auguste. *Cours de philosophie positive*, 6 tom. (Paris, 1830/42).

Cook, Harold. 'From the Scientific Revolution to the Germ Theory.' In Loudon (1997).

Cooter, Roger and John Pickstone. *Companion to Medicine in the Twentieth Century.* (London & New York: Routledge, 2003).

Coulter, Harris. http://www.pnc.com.au/~cafmr/online/research/index.html.

Dawkins, Richard. *The Selfish Gene* (Oxford: Oxford University Press, 1976).

———. *The Blind Watchmaker.* (New York: Norton & Company, Inc., 1986).

Descartes, René. 'Discourse on the Method', in *The Philosophical Writings of Descartes,* translated by John Cottingham, Robert Stoothoff and Dugald Murdoch. Vol. 1 (Cambridge: Cambridge University Press, 1992).

———. *Philosophical Essays and Correspondence.* Ed. Roger Ariew. (Indianapolis: Hackett Publishers, 2000).

———. *Rules for the Direction of the Mind,* 1628; http://faculty.uccb.ns.ca/ philosophy/kbryson/rulesfor.htm.

Diamond, Jared. *Guns, Germs and Steel.* (New York: W W Norton & Company, 1999).

Dickens, William and James Flynn, 'Great Leap Forward', *New Scientist,* April 21, 2001.

Dijksterhuis, E.J. *The Mechanization of the World Picture.* Translated by C. Dikshoorn. (Oxford: Clarendon Press, 1961).

Duke, Martin. *The Development of Medical Techniques and Treatments: From Leeches to Heart Surgery.* (Madison, CT: International Universities Press., Inc., 1991).

Elwood, J. Mark. *Causal Relationships in Medicine: A Practical System for Critical Appraisal.* (Oxford: Oxford University Press, 1988).

Evans, Alfred S. *Causation and Disease: A Chronological Journey.* (New York: Plenum, 1993).

Evans, Dylan. *Placebo: Mind over Matter in Modern Medicine.* (Oxford University Press, 2004).

Evans, Imogen, Hazel Thornton & Iain Chalmers. *Testing Treatments: Better Research for Better Healthcare.* (London: British Library, 2007) http://www.jameslindlibrary.org/pdf/testing-treatments.pdf

Faber, Knud Helge. *Nosography: The Evolution of Modern Internal Clinical Medicine.* (New York: AMS Press, 1978).

Foss, Lawrence. 'The Nobel Prize and the Biomedical Paradigm: Is It Time For A Change', *Theoretical Medicine and Bioethics* 19 (1998) 621–44.

French, Roger. 'Harvey in Holland: Circulation and the Calvanists', in Roger French and Andrew Wear. Eds. *The Medical Revolution of the Seventeenth Century.* (New York: Cambridge University Press, 1989).

———. 'The Anatomical Tradition', in Bynum and Porter (1993b).

———. *Medicine Before Science: The Rational and Learned Doctor from the Middle Ages to the Enlightenment.* (Cambridge: Cambridge University Press, 2003).

Fry, Stephen. 'In Search of the Planet's Most Endangered Species: Retracing the Wild Quest of His Friend Douglas Adams', *The Guardian, G2,* 02/09/2009.

Gaines, Atwood D. and Robbie Davis-Floyd. 'On Biomedicine', in *Encyclopedia of Medical Anthropology,* edited by Carol and Melvin Ember. (Yale: Human Relations Area Files, 2003). http://www.davis-floyd.com/userfiles/Biomedicine.pdf

Golub, Edward S. *the Limits of Medicine: How Science Shapes Our Hope for the Cure.* (Chicago: Chicago University Press, 1997).

Goodman, Jordan. 'Pharmaceutical Industry', in Cooter & Pickstone (2003).

———. *Useful Bodies.* (Baltimore: John Hopkins University Press, 2003).

Guyatt, David. 'Unit 731: Military Logs Light Military Research Military Fires' (1997), http://www.copi.com/articles/guyatt/unit_731.html

Hamilton, Leonard W. and C. Robin Timmons. 'Psychopharmacology', in A. Colman, ed. *Companion Encyclopaedia of Psychology, Vol.1.* (London: Routledge, 1994); http://www.rci.rutgers.edu/~lwh/drugs/psypharm.htm.

Hanfling, Oswald. *Cause and Effect.* (Milton Keynes: Open University Press, 1980).

Harris, Sheldon H. *Factories of Death.* (London: Routledge, 1995).

Hart, H.L.A. and A. Honoré. *Causation in the Law.* (Oxford: Clarendon Press, 1959).

Hawkins, D.F. *Drugs and Pregnancy: Human Teratogenesis and Other Problems.* (Edinburgh: Churchill Livingstone, 1983).

Healy, David. *The Creation of Pharmacology.* (Cambridge, MA: Harvard University Press, 2002).

——. *The Psychiatric Drugs Explained.* 5th Edition. (Edinburgh: Churchill Livingstone/Elsevier, 2009).

Heidegger, Martin. 'The Question Concerning Technology', in *The Question Concerning Technology and Other Essays*, translated by William Lovitt (New York: Harper & Row Ltd., 1982).

Hennekens, C.H. and J.E. Buring (1987). *Epidemiology in Medicine.* (Boston: Little Brown, 1993).

Hesslow, Germund. 'Do We Need A Concept of Disease?' *Theoretical Medicine,* 14,1(1993); http://www.springerlink.com/content/u4754t02353r3371/ full-text. pdf.

Hirst, Leonard Fabian. *Conquest of Plague: A Study of the Evolution of Epidemiology.* (Oxford: Clarendon Press, 1953).

Hirst, P.Q. *Durkheim, Bernard and Epistemology.* (London: Routledge & Kegan Paul, 1975).

Hoefer, Carl. 'Causal Determinism'. *Stanford Encyclopedia of Philosophy*: http://plato.stanford.edu/entries/determinism-causal/ (first published January 23, 2003; substantial revision January 21, 2010).

Hrobjartsson, A. and P.C. Gotsche. 'Is the Placebo Powerless? An Analysis of clinical Trails Comparing Placebo with No Treatment', *New England Journal of Medicine* 344 (2001) 1594–602.

Hudson, Robert P. *Disease and Its Control: The Shaping of Modern Thought.* (Westport, CT: Greenwood Press, 1983).

Hume, David. *Dialogues Concerning Natural Religion (1779)*, ed. Richard Popkin. (Indianapolis: Hackett, 1998).

Jain, Kewal K. *The Handbook of Nanomedicine.* (Totowa, NJ: Humana Press, 2008); http://www.springerlink. com/content/k653k7/.

Jensen, Uffe Juul. 'A Critique of Essentialism in Medicine', in *Health, Disease, and Causal Explanations in Medicine.* Eds. Lennart Nordenfelt. (Dordrecht: D Reidel Publishing Company, 1984).

Jonas, Hans. *The Phenomenon of Life: Toward a Philosophical Biology* (New York: Harper and Row, 1966).

Judson, A.B. 'The Human Body Viewed as a Machine', *New York Medical Journal,* July 24, 1909.

Kaptchuk, Ted A. 'Powerful placebo: the dark side of the randomised controlled trial', *The Lancet,* 351(1998) 1722–25.

Kennington, Kenneth. 'Descartes and the Mastery of Nature', in Spicker (1978).

King, Lester S. *The Growth of Medical Thought.* (Chicago: University of Chicago Press, 1963).

———. *The Philosophy of Medicine: The Early Eighteenth Century.* (Cambridge, MA & London, 1978).

———. *Medical Thinking: A Historical Preface.* (Princeton, NJ: Princeton University Press, 1982).

Kiple, Kenneth. 'The History of Disease', in Porter (1996).

Kleinman, Arthur. 'What is Specific to Western Medicine?' in Bynum and Porter, Vol. 1(1993).

Kloppenburg, Jack Ralph, Jr. *First The Seed: The Political Economy of Plant Technology, 1492–2000* (Cambridge and New York: Cambridge University Press, 1990).

Koshi, Edvin B., Christine Ann Short. 'Placebo Theory and Its Implications for Research and Clinical Practice: A Review of Recent Literature', *Pain Practice* Volume 7, Issue 1, (2007) 4–20; http://www.wired.com/medtech/drugs/magazine/17-09/ff_placebo_effect? currentPage=all.

Kuhn, T.S. *The Structure of Scientific Revolutions.* (Chicago: Chicago University Press, 1970).

Lakatos, Imre. 'Falsification and the Methodology of Scientific Research Programmes', in *Criticism and the Growth of Knowledge: Proceedings of the International Colloquium in the Philosophy of Science*, edited by Lakatos and Musgrave. (Cambridge: Cambridge University Press, 1970).

Lasagna, Louis. 'The Pharmaceutical Revolution: Its Impact on Science and Society', *Science* 166(1969)1227–38).

Lawrence, Ghislaine. 'Surgery (Traditional)', in Bynum and Porter, Vol. 2. (London: Routledge, 1993).

———. *Technologies of Modern Medicine.* (London: Science Museum, 1994).

Lee, Keekok. 'Kuhn – A Re-appraisal', *Explorations in Knowledge* 1,1(1984) 33–88.

———. *A New Basis for Moral Philosophy.* (London: Routledge and Kegan Paul, 1985).

———. *The Positivist Science of Law.* (Hants: Avebury, 1989a).

———. *Social Philosophy and Ecological Scarcity.* (Routledge: London, 1989b).

———. 'Designer Mountains: The Ethics of Nanotechnology', *Terra Nova: Nature and Culture*, 2(1997) 127–36.

———. *The Natural and the Artefactual: Implications of Deep Science and Deep Technology for Environmental Philosophy.* (Lanham: Lexington Books, 1999).

———. *Philosophy and Revolutions in Genetics: Deep Science and Deep Technology*, (Palgrave: Basingstoke, 2005, 2nd edition).

———. *Zoos: A Philosophical Tour.* (London: Palgrave MacMillan, 2006).

———. 'Homo Faber: The Unity of the History of Technology and the Philosophy of Technology', in *New Waves in Philosophy of Technology*, edited by Jan-Kyrre Berg Olsen, Evan Selinger and Søren Riis. (London: Palgrave MacMillan, 2008).

———. 'Towards Constructing Post-postmodern 21st Century Sciences: The Relevance of Classical Chinese Medicine', in *Asia/Europe Dialogue and the Making of Modern Science*, edited by Arun Bala. (Basingstoke & Singapore: Palgrave Macmillan and the Institute of South East Asian Studies, 2011).

Liberman, R. 'An Experimental Study of the Placebo Response Under Three Different Situations of Pain', *Journal of Psychiatric Research*, 2, 223–46.

Lilienfield, Abraham, M. *Times, Places and Persons. Aspects of the History of Epidemiology.* (Baltimore: John Hopkins University Press,1980).

Lock, Margaret, Allan Young and Alberto Cambrosio. Eds. *Living and Working with the New Medical Technologies.* (Cambridge: Cambridge University Press, 2000).

Lock, Stephen. 'The randomised controlled trial – a British invention', in Lawrence (1994).

——. 'Medicine in the Second Half of the Twentieth Century', in Loudon (1997).

Loudon, Irvine. *An Illustrated History of Western Medicine.* (Oxford: OUP, 1997).

Loux, Françoise. 'Folk Medicine', in Bynum and Porter. Eds. (London: Routledge, 1993a).

Löwy, Llana. 'Recent historiography of biomedical research', in Lawrence (1994).

——. 'Trustworthy knowledge and desperate patients: clinical tests for new drugs from cancer to AIDS', in Lock et al. (2000).

Macartney, Fergus J. 'Diagnostic Logic', in Phillips (1989).

Magner, Lois N. *A History of Medicine.* (New York: Marcel Dekker, Inc., 1992).

Mackie, J.L. *The Cement of the Universe: A Study of Causation.* (Oxford: Clarendon Press, 1974).

McInstosh, Robert P. *The Background of Ecology: Concept and Theory.* (Cambridge: Cambridge University Press, 1985).

Marchant, Joanna. 'Know Your Enemy', *New Scientist,* 168 (4 November 2000) 46–50.

Marmot, Michael. 'Historical perspective: the social determinants of disease – some blossoms', http://www.epi-perspectives. com/content/2/1/4.

Maturana, Humberto R. and Francisco J. Varela. *Autopoiesis and Cognition: The Realization of the Living* (Dordrecht/Boston : D. Reidel Publishing Company, 1980).

——. *The Tree of Knowledge: The Biological Roots of Human Understanding* (Boston: Shambhala, 1988).

Maturana, Humberto R., Francisco J. Varela and R. Uribe. 'Autopoiesis: The Organization of Living Systems', *Biosystems,* 5 (1974) 187–96.

Maulitz, Russell C. 'The Pathological Tradition', in Bynum & Porter, Vol. 1. (London: Routledge, 1993).

Maynard, Alan and Iain Chalmers. Editors. *Non-random Reflections on Health Services, on the 25th Anniversary of Archie Cochrane's* Effectiveness and Efficiency. (London: BMJ, 1997).

Mayr, Ernst. *The Growth of Biological Thought* (Cambridge, MA: Belknap Press of Harvard University Press, 1982).

Mazzarello, Paolo. "A Unifying Concept: the history of cell theory." In *Nature Cell Biology* 1, E13–E15, 1999. http://www.nature.com/ncb/journal/ v1/n1/ full/ ncb0599_E13.html.

McKeown, Thomas. *The Role of Medicine: Dream, Mirage or Nemesis?* (Oxford: Blackwell, 1979).

McKie, Robert. 'Medicine Man at the $100m Biotech Temple', *The Observer, Business Section,* (9 October 1994) 6.

——. 'Revealed: The Secret of Human Behaviour', *The Observer* (11 February 2001) 1.

Meek, James. 'Sons Created to Beat Blood Disease', *The Guardian* (17 October 2000).

Mendel, Gregor. *Experiments in Plant Hybridization* edited by J.H. Bennet (Edinburgh: Oliver and Boyd, 1965).

Merchant, Carolyn. *The Death of Nature: Women, Ecology and the Scientific Revolution.* (San Francisco: Harper & Row, Publishers, 1980).

Metchnikoff, Elie. *The Nature of Man: Studies in Optimistic Philosophy.* (London, Heinemann, 1903).

Metchnikoff, Elie. *The Founders of Modern Medicine: Pasteur, Koch,* Lister. (New York: Walden Publications, 1939).

Mettrie, Julian Offray de la. *Man A Machine.* 1748; in English in 1750, http://www. cscs.umich.edu/~crshalizi/ LaMettrie/ Machine/

Mez-Mangold, Lydia. *A History of Drugs.* (New York; Barnes & Noble, 1986).

Mill, J.S. *A System of Logic.* 1843 (Toronto and London: University of Toronto Press, Routledge and Kegan Paul, 1974): http://www.gutenberg.org/etext/27942.

Mitcham, Carl. *Thinking Through Technology: The Path between Engineering and Philosophy.* (Chicago: Chicago University Press, 1994).

Moerman, Daniel E. *Meaning, Medicine and the 'Placebo Effect.'* (Cambridge: Cambridge University Press, 2002); http:// catdir.loc.gov/catdir/samples/ cam033/ 2002020167.pdf

Mol, Annemarie. 'Pathology and the clinic: an ethnographic presentation of two atheroscleroses', in Lock et al. (2000).

——. *The Body Multiple: Ontology in Medical Practice.* (Durham, N.C.: Duke University Press, 2002).

Müller, Ulrich, Paul C. Fletcher and Holger Steinberg. 'The Origin of Pharmacopsychology: Emil Kraepelin's Experiments in Leipzig, Dorpat and Heidelberg (1882–1992)', *Psychopharmacology* (Berl), 2006, 184: 131–8.

Mumford, Lewis. *Technics and Civilization* (London: George Routledge & Sons, Ltd., 1946).

——. *The Myth of the Machine: Technics and Human Development* (London: Secker and Warburg, 1967).

Murphy, Edmond A. *The Logic of Medicine.* (Baltimore, MD: John Hopkins University Press, 1976, 1997).

Nagl, Sylvia. *Rethinking Causation for Complex Systems in Biomedicine: Challenges and New Approaches.* http://www.kent.ac.uk/secl/philosophy/ jw/2006/ capim/ Nagl.pdf.

National Research Council. *Genetic Engineering of Plants* (Washington, DC: National Academy Press, 1984).

Nicolson, Malcolm. 'The Art of Diagnosis: Medicine and the Five Senses', in Bynum and Porter, Vol. 2 (London: Routledge, 1993a).

Niemi, Maj-Britt. 'Placebo Effect: A Case in the Mind', *Scientific American Mind* February 2009.

Nutton, Vivian. 'The Rise of Medicine', in Roy Porter (1996).

Olby, Robert C. *Origins of Mendelism* (London: Constable, 1966).

Pelling, Margaret. 'Contagion, Germ Theory, Specificity', in Bynum and Porter, Vol. 1. (London: Routledge, 1993a).

O'Leary, K. Daniel and Thomas D. Borkovec. 'Conceptual, Methodological, and Ethical Problems of Placebo Groups in Psychotherapy Research', *American Psychologist*, (September 1978) 821.

O'Malley, C.D. and J.B. de C.M. Saunders. Eds. *Leonardo da Vinci on the Human Body.* (New York: Schuman, 1952).

Osborne, David K. 'Greek Medicine', 2009, http://www.greekmedicine. net/ Principles_ of_ Treatment/ Introduction_to_Therapeutics_in_Greek_ Medicine.html.

Pagel, Walter. *Parecelsus: An Introduction to Philosophical Medicine in the Era of the Renaissance.* (Basel: S. Karger, 1958).

Paley, William, 1802, *Natural Theology* (Indianapolis: Bobbs-Merrill, 1963).

Pasteur, Louis. 'Claude Bernard: idée de l'importance de ses travaux, de son enseignement et de sa méthode', *Moniteur Universel*, No 311 (7th novembre 1866) 1284–85.

Parascandola, M. 'Epidemiology: Second-Rate Science?', *Public Health Reports*, 113 (1998) 312–20.

Perrine, Daniel M. *The Chemistry of Mind-altering Drugs: History of Pharmacology and Cultural Context.* (New York: Oxford University Press, 1996).

Petrovic, P., Eija Kalso, Karl Magnus Petersson and Martin Ingvar 'Placebo and Opiod Analgesia: Imaging A Shared Neuronal Network." *Science*, 295, 5560 (March 2002) 1737–40.

Phillips, Calbert. *Logic in Medicine.* (London: British Medical Journal, 1989).

Pickstone, John (ed). *Medical Innovations in Historical Perspective.* (London: Macmillan, 1992).

Pisek, Paul E. and Trisha Greenhalgh. 'The Challenge of Complexity in Health Care', BMJ September 2001; 323 (2001) 625–8. http://www.pubmedcentral. nih.gov/ articlerender.fcgi?artid=1121189.

Popper, Karl. *The Logic of Scientific Discovery.* (London: Routledge, 1959).

——. *Conjectures and Refutations: The Growth of Scientific Knowledge.* (London: Routledge, 1963).

Porter, Roy. 'The rise of physical examination', in Bynum and Porter (1993b).

——. *The Greatest Benefit to Mankind: A Medical History of Humanity from Antiquity to the Present.* (London: Fontana, 1999).

——. Ed. *The Cambridge Illustrated History of Medicine.* (Cambridge: Cambridge University Press, 1996).

——. 'What is Disease?' in Porter (1996).

——. 'Medical Science' in Porter (1996).

——. 'Hospitals and Surgery' in Porter (1996).

——. 'The Eighteenth Century' in *The Western Medical Tradition: 800 BC to AD 1800*, edited by Lawrence I. Conrad et al. (Cambridge: Cambridge University Press, 2003).

Reiser, Stanley Joel. *Medicine and the Reign of Technology.* (Cambridge: Cambridge University Press, 1990).

——. 'The Science of Diagnosis: Diagnostic Technology', in Bynum & Porter, Vol. 2 (1993a).

——. 'Technology and the Use of the Senses in Twentieth-century medicine', in Bynum and Porter (1993b).

——.Resnek, Lawrie. *The Nature of Disease.* (London: Routledge & Kegan Paul, 1987).

René Spiegel. *Pharmacology: An Introduction.* (New York: Wiley, John & Sons, 2003).

Reynolds, G.S. *A Primer of Operant conditioning.* (Glenview, IL: Scott, Foresman & Co., 1975).

Rheinberger, Hans-Jörg. 'Beyond nature and culture: modes of reasoning in the age of molecular biology and medicine', in Lock et al. (2000).

Riese, Walther. *The Conception of Disease, its History, its Versions and its Nature.* (New York: Philosophical Library, 1953).

——. *'The Philosophical Presuppositions of Present-day Medicine', Bulletin of the History of Medicine,* 30(1956) 163–74.

Rindos, David. *The Origins of Agriculture: An Evolutionary Perspective* (Orlando and London: Academic Press, Inc., 1984).

Rosen, Robert. 'Organisms as Causal Systems Which Are not Mechanisms: An Essay into the Nature of Complexity', in I.W. Richardson. *Theoretical Biology and Complexity: Three Essays on the Natural Philosophy of Complex Systems.* (London: Academic Press, 1985).

Sample, Ian. 'Our flexible friend', *The Guardian,* 7 April 2009.

Schaffner, Kenneth F. *Discovery and Explanation in Biology and Medicine.* (Chicago and London: Chicago University Press, 1993).

Shapiro, A.K. 'Placebo Effects in Medicine, Psychotherapy, and Psychoanalysis', in *Handbook of Psychotherapy and Behaviour Change,* edited by A.E. Bergin & S.L. Garfield. (New York, Wiley, 1971).

Schardein, J.L. *Drugs as Teratogens.* (Cleveland, OH: CRS Press, 1975).

Seigworth, Gilbert R. MD. 'Bloodletting Over the Centuries', *New York State Journal of Medicine,* (December 1980) 2022–8.

Sharpe, Robert. *The Cruel Deception: The Use of Animals in Medical Research.* (Wellinborough: Thorsons Publishing Group, 1988).

Shorter, Edward. 'Primary Care', in Porter (1996).

Silverman, WA. *Human Experimentation: a Guided Step into the Unknown.* (Oxford: Oxford University Press, 1985).

——. *'Where's the Evidence?'* in *Controversies in Modern Medicine.* (Oxford: Oxford University Press, 1998).

Simmonds, Norman W. *Principles of Crop Improvement* (New York: Longman, 1979).

Sober, Elliott and Richard Lewontin. 'Artifact, Cause and Genic Selection', *Philosophy of Science,* 49(1982) 157–80.

Sperati, G. 'Craniotomy through the ages', http://www.ncbi.nlm.nih.gov/ pmc/ articles/PMC2640049/.

Spicker, F. Stuart. *Organism, Medicine, and Metaphysics.* (Boston: D Reidel, 1978).

Spicker, F. Stuart and H. Trisram Engelhardt, Jr. *Philosophical Dimensions of the Neuro-Medical Sciences.* (Dordrecht/Boston: D Reidel Publishing Company, 1976).

Stein, Dan. J. (Ed). *Philosophy of Psychopharmacology.* (New York: Cambridge University Press, 2008).

Strawson, P.F. *Individuals: An Essay in Descriptive Metaphysics.* (London: Methuen, 1959).

Susser, M. *Causal Thinking in the Health Services: Concepts and Strategies in Epidemiology.* (London: Oxford University Press, 1973).

——. 'What is a cause and how do we know one? A grammar for pragmatic epidemiology', *American Journal Epidemiology* 133 (1991) 635–48.

Tansey, E.M. 'The Physiological Tradition', in Bynum and Porter, Vol.1 (1993a).

——.'From the Germ Theory to 1945', in *Western Medicine: An Illustrated History.* (Oxford: Oxford University Press, 1997).

Taylor, F. Kräupl. *The Concepts of Illness, Disease and Morbus*. (Cambridge: Cambridge University Press, 1979).

Temkin, Owsei. 'The Role of Surgery in the Rise of Modern Medical Thought' and 'Metaphors of Human Biology', in *The Double Face of Janus*. (Baltimore and London: the Johns Hopkins University Press, 1977).

——. *'On Second Thought and Other Essays', in the History of Medicine and Science'* (Baltimore and London: The Johns Hopkins University, 2002).

Thagard, Paul. *Concept of Disease: Structure and Change* (Waterloo University, 1997); http://watarts.uwaterloo.ca/~pthagard/Articles/Pages/Concept.html

——. *How Scientists Explain Disease*. (Princeton, NJ: Princeton University Press, 2000).

Thornton, Stephen. Entry on Popper in *Stanford Encyclopedia of Philosophy*, 2009. http://plato.stanford.edu/entries/popper/

Tröhler, Ulrich. 'Surgery (Modern)', in Bynum and Porter Vol. 2 (1993a).

Varela, Francisco J. *Principles of Biological Autonomy* (New York and Oxford: North Holland, 1979).

Veltman, Kim H. 2011 *Leonardo da Vinci: Studies of the Human Body and Principles of Anatomy*. http://www.mmi.unimaas.nl/people/Veltman/articles/leonardo/Lenardo%20da%20Vinci%20Studies%20of%20the%20Human%20Nody%20and%20Prinicples%20of%20Anatomy.html

Waller, John. *The Discovery of the Germ*. (Cambridge: IconBooks UK, 2004).

Walsh, B.T., S.N. Seidman, R. Sysko and M. Gould 'Placebo Response in Studies of Major Depression', *Journal of American Medical Association* 287 (2002) 1840–7.

Wangensteen, Owen H. and Sarah D. Wangesnsteen. *The Rise of Surgery. From Empiric Craft to Scientific Medicine*. (Mineapolis, MN: University of Minnesota Press, 1978).

Watts, Geoff. 'Looking to the Future', in Porter (1996).

Wear, Andrew. 'Medical Practice in Late Seventeenth- and Early Eighteenth-century England: Continuity and Union', in *The Medical Revolution of the Seventeenth Century*, edited by Andrew Wear and Roger French (Cambridge: Cambridge University Press, 1989).

Wellman, Kathleen Anne. *La Mettrie: medicine, philosophy, and enlightenment*. (Durham: Duke University Press, 1992).

Westfall, Richard. *The Construction of Modern Science: Mechanisms and Mechanics*. (Cambridge: Cambridge University Press, 1977).

Wetherall, Miles. 'Drug Treatment and the Rise in Pharmacology', in Porter (1996).

Wilkinson, Lisa. 'Epidemiology', in Bynum and Porter, Vol. 2 (1993a).

Winslow, Charles-Edward Amory. *The Conquest of Epidemic Disease. A Chapter in the History of Ideas*. (Madison, WI: University of Wisconsin press, 1980).

Wolf, Jonathan. 'The secret of good health? Don't get ill in the first place', *The Guardian*, 7 April 2009.

Wolff, Manfred E. Editor of the Fifth Edition of *Burger's Medicinal Chemistry and Drug Discovery*. (New York and Chichester: John Wiley & Sons, Inc., 1995).

Wootton, David. *Bad Medicine*. (Oxford: Oxford University Press, 2006).

Worboys, Michael. *Spreading Germs: Disease Theories and Medical Practice in Britain, 1865–1900*. (Cambridge: Cambridge University Press, 2000).

Wulff, Henrik R. 'Philosophy of Medicine: From a Medical Perspective', *Theoretical Medicine and Bioethics*, 13 (1992).

——. 'The Causal Basis of the Current Disease Classification' in *Health, Disease, and Causal Explanations in Medicine*, edited by Lennart Nordenfelt. (Dordrecht: D Reidel Publishing Company, 1984).

Wulff, Henrik R and Peter C. Gotzsche. *Rational Diagnosis and Treatment: Evidence-Based Clinical Decision-Making*, Third Edition (Oxford: Blackwell Science, 2000).

Wulff, Henrik R, Stig Andur Pedersen and Raben Rosenberg. *Philosophy of Medicine: An Introduction* (Oxford: Blackwell, 1990).

Wulff, Henrik, R. and Morten Skydsgaard. 'The Evolution of Western Medicine', *Medicine, Health Care and Philosophy*, 1 (1998).

Wynder and Hoffman. 'Smoking and Lung Cancer: Scientific Challenges and Opportunities '(1994); http://cancerres.aacrjournals.org/cgi/reprint/54/20/5284.

Yates, F.E. 'Molecules and Modern Medicine: Where is the Patient?' *QJ Med*, 88 (1995) 88: 69–72; http://qjmed.oxfordjournals.org/cgi/reprint/88/1/69.pdf

——. 'Self-organizing Systems.', in *The Logic of Life: The Challenge of Integrative Physiology*, edited by C.A.R. Boyd and D. Noble. (Oxford: Oxford University Press, 1993).

Zimmerman, Michael. *Heidegger's Confrontation with Modernity: Technology, Politics, Art* (Bloomington: Indiana University Press, 1990).

Index

CPSIA information can be obtained at www.ICGtesting.com
Printed in the USA
LVOW04*2141050815

448976LV00011B/154/P